ADVANCES IN
MOLECULAR AND
CELL BIOLOGY

Volume 25 • 1998

OXYRADICALS IN MOLECULAR BIOLOGY

ADVANCES IN MOLECULAR AND CELL BIOLOGY

OXYRADICALS IN MEDICAL BIOLOGY

Series Editor E. EDWARD BITTAR
Department of Physiology
University of Wisconsin
Madison, Wisconsin

Guest Editor JOE M. McCORD
Webb-Waring Institute
University of Colorado Health Sciences Center
Denver, Colorado

VOLUME 25 • 1998

JAI PRESS INC.

Greenwich, Connecticut *London, England*

55 Old Post Road No. 2
Greenwich, Connecticut 06836

JAI PRESS LTD.
38 Tavistock Street
Covent Garden
London WC2E 7PB
England

ISBN: 0-7623-0379-4

Transferred to digital printing 2006

CONTENTS

LIST OF CONTRIBUTORS

Patrick A. Baeuerle	Tularik Inc. San Francisco, California
Bruce A. Freeman	Department of Anesthesiology The University of Alabama at Birmingham Birmingham, Alabama
Sonia C. Flores	Webb-Waring Institute University of Colorado Health Sciences Center Denver, Colorado
Irwin Fridovich	Department of Biochemistry Duke University Medical Center Durham, North Carolina
D. Neil Granger	Department of Physiology LSUMC School of Medicine Shreveport, Louisiana
Norman R. Harris	Department of Physiology LSUMC School of Medicine Shreveport, Louisiana
Brooks M. Hybertson	Webb-Waring Institute University of Colorado Health Sciences Center Denver, Colorado
Jonathan A. Leff	Merck and Co., Inc. Rahway, New Jersey
Joe M. McCord	Webb-Waring Institute University of Colorado Health Sciences Center Denver, Colorado

Sally K. Nelson	Webb-Waring Institute University of Colorado Health Sciences Center Denver, Colorado
Larry W. Oberley	Radiation Research Laboratory University of Iowa Iowa City, Iowa
Terry D. Oberley	Pathology Service William S. Middleton Memorial Veterans Hospital Madison, Wisconsin
Klaus Schulze-Osthoff	Department of Internal Medicine I University of Tübingen Tübingen, Germany
David Patterson	Eleanor Roosevelt Institute Denver, Colorado
John E. Repine	Webb-Waring Institute University of Colorado Health Sciences Center Denver, Colorado
Homero Rubbo	Department of Anesthesiology The University of Alabama at Birmingham Birmingham, Alabama

PREFACE

While we scientists would like to think that our view of the world is purely objective, it is in fact colored by preconceived notions, false assumptions, and prejudices that are not so different from those that distort perceptions for the rest of humankind. (Perhaps we can claim that we are at least *more* objective than they are.) In any case, the rapid expansion of the area of free radical biology in the last 25 years has occurred within a framework of assumptions and preconceived notions that has at times directed the course of this movement. The most dominant of these notions has been the view that free radical production is without exception a bad thing, and that the more efficient our elimination of these toxic substances, the better off we will be. The very important observation by Bernard Babior and colleagues in 1973 that activated phagocytes produce superoxide in order to kill microorganisms, served to illustrate that constructive roles are possible for free radicals. For many in the field, however, this merely underscored the deadly nature of oxygen-derived radicals, both from the microbe's point of view and from the host's as well. (Phagocyte-produced superoxide is responsible in part for the tissue injury manifested as inflammation. See Harris and Granger, Chapter 5, and Leff, Hybertson and Repine, Chapter 6.)

Mother Nature, however, has a penchant for being able to make a silk purse from a sow's ear. If one is dealt a bad hand, one must simply make the best of it. After two decades of focusing on the destructive side of free radicals, the last few years have begun to reveal a new and finer perspective on free radical metabolism—a role in regulation of cellular function (see Schulze-Osthoff and Baeuerle, Chapter 2). Evidence from a number of sources suggests that an increase in the oxidative status of a cell encourages that cell to grow and divide. Increasing the expression of manganese superoxide dismutase can suppress the malignant phenotype of melanoma cells (see Oberley and Oberley, Chapter 3). Oxidative stress beyond a certain point seems to trigger a process in normal cells known as programmed cell death or *apoptosis* (from the Greek, literally "to fall apart"). Is this suicide response an evolutionary fail-safe device to curtail tumorogenesis? Does oxidative stress-induced apoptosis account for the loss of immune cells in AIDS (see Flores and McCord, Chaper 4)?

This volume attempts to present the spectrum of roles, both good and bad, played by active oxygen species as understood at this point in the evolution of the field of free radical biology. Admittedly and obviously, our knowledge is incomplete, and still rapidly expanding.

Joe M. McCord
Guest Editor

REFERENCE

Babior, B.M., Kipnes, R.S., & Curnutte, J.T. (1973). Biological defense mechanisms. The production by leukocytes of superoxide, a potential bactericidal agent. J. Clin. Invest., 52, 741-744.

AN OVERVIEW OF OXYRADICALS IN MEDICAL BIOLOGY

Irwin Fridovich

Advances in Molecular and Cellular Biology
Volume 25, pages 1-14.

ISBN: 0-7623-0379-4

I. INTRODUCTION–OXYRADICALS IN LIVING CELLS?

Knowledge of the normal, or physiological, provides an excellent foundation for building an understanding of the abnormal, or pathological. It is therefore appropriate that a book on Oxyradicals in Medical Biology begin with an overview of the normal biology of oxygen-derived free radicals. That there is a normal biology of oxyradicals would have seemed absurd to an earlier generation of scientists to whom oxyradicals were products of radiolysis and of gas phase reactions, with no relevance to ordinary biology.

This view has been completely changed by explorations conducted during the past three decades. We now know that oxyradicals, such as superoxide anion, $O_2^{\cdot -}$, are routinely produced within and upon living cells and that this occurs during both spontaneous and enzyme-catalyzed reductions of molecular oxygen. We also know that O_2^{-} is itself capable of damaging essential biomolecules and that it can, moreover, engender species more reactive than itself, among which are $HO_2^{\cdot}$, Fe(II)O, and HO·.

This witches' brew of oxygen-derived free radicals would make aerobic life virtually impossible were it not for a multilayered system of defenses. Among these defenses are enzymes which selectively eliminate $O_2^{\cdot -}$, H_2O_2, and alkyl hydroperoxides; low molecular weight antioxidants, both water soluble for action in the cytosol and lipid soluble for action within membranes, which limit the extent of free radical chain reactions; enzymes which repair oxidatively damaged DNA and which recycle oxidatively damaged proteins; and finally oxidation-resistant isoenzymes which, under conditions of oxidative stress, replace their oxidation-sensitive analogues.

Let us now examine the biology of oxygen-derived free radicals which underlies the toxicity of molecular oxygen, the oxygen-dependent toxicities of many compounds, and numerous pathologies. Then, having looked at the problems that accompany the aerobic lifestyle, we will enumerate the defenses that allow us to live in the oxygenated biosphere bequeathed to us by the photosynthetic organisms. All of this, the problems posed by oxyradicals, and the defenses, will serve as an entrée to the chapters that follow.

II. MOLECULAR OXYGEN AND THE SPIN RESTRICTION

O_2 in the ground state is paramagnetic, which denotes unpaired electronic spins. Indeed, O_2 contains two unpaired electrons with parallel spin states. This electronic structure imposes a kinetic barrier towards divalent reduction. This hindrance is due to the relative slowness of the inversion of electronic spins and to the need for such inversion of spin when an electron pair is inserted into $O_2^{\cdot}$. This can be made clear by using vertical arrows to depict electronic spin states, as in reaction 1 that depicts the divalent reduction of O_2 to H_2O_2 :

$$\underbrace{\uparrow\uparrow}_{O_2} + \underbrace{\uparrow\downarrow}_{e\ \text{pair}} \xrightarrow{+2H^+} \underbrace{\uparrow\downarrow\ \uparrow\downarrow}_{H_2O_2} \qquad (1)$$

We see three up-pointing arrows among the reactants on the left and only two in the product on the right. Obviously one electronic spin must be inverted during this divalent reduction and that imposes the kinetic bottleneck that is often called the spin restriction.

The spin restriction is a formidable barrier to reaction because electron transfer must occur during the moment of collision and because the time required for inversion of electronic spin is much longer than the lifetime of a collisional complex. Since there is relatively much more time between collisions, it is easier to reduce O_2 by a univalent pathway, in which electrons are donated to it one at a time. This allows for spin inversion to occur during the long times between collisions, rather than requiring that it occur during the fleeting lifetime of the collisional complex.

The spin restriction thus makes O_2 less reactive than it would otherwise be and at the same time favors the univalent pathway. The first of these consequences is fortunate in that it prevents the spontaneous oxidation of many organic compounds. The second forces living things to deal with the oxyradicals that are intermediates on the univalent pathway, and this is surely the lesser of two evils. That it is so is demonstrated by photodynamic effects in which a dye mediates excitation of O_2 by light. Excitation of O_2 by ~23 kcal inverts one electronic spin and the resultant singlet oxygen can rapidly attack many compounds including the unsaturated fatty acid residues found in biological membranes. Houseflies that have ingested methylene blue-laced sugar water are rapidly killed by light, due to such a photodynamic effect.

III. O_2^{-}—HOW MUCH?

The univalent pathway involves three intermediates between O_2 and H_2O and these are O_2^-, H_2O_2 and $HO\cdot$. The first of these, O_2^-, is easily made both *in vitro* and *in vivo*. This is not to say that O_2^- is ordinarily a major product of biological oxygen reduction. To the contrary, most reduction of O_2 in respiring cells is catalyzed by cytochrome oxidase which manages an overall tetravalent reduction of O_2 to $2H_2O$, without the release of any intermediates. Respiring cells try to minimize O_2^- production and do so quite successfully. O_2^- production is consequently only a small percentage of O_2 reduction. How much O_2^- is made in respiring cells?

Escherichia coli, growing in a rich aerobic medium at 37°0C, has been examined (Imlay and Fridovich, 1991). This cell uses 6.2×10^6 molecules of O_2 per second and produces 8×10^3 O_2^- in the same time. Were O_2^- perfectly stable, its concentration within an *E. coli* cell would increase at a rate of 4.2 μM per second! Since a hepatocyte and *E. coli* possess comparable defenses against O_2^-, it seems reasonable to suppose that O_2^- is made at comparable rates in these very different cells.

IV. O_2^-—SOME PROPERTIES

O_2^- can act either as a reductant or as an oxidant. Frequently used assays have been based on both of these properties. There are thus assays in which O_2^- reduces nitroblue tetrazolium (Beauchamp and Fridovich, 1971), or cytochrome *c* (McCord and Fridovich, 1969), and others in which it oxidizes epinephrine (McCord and Fridovich, 1969), 6-hydroxydopamine (Heikkila and Cabbat, 1976), hydroxylamine (Elstner and Heupel, 1976), or pyrogallol (Marklund and Marklund, 1974). Indeed, in the dismutation reaction, b, O_2^- acts both as a reductant and as an oxidant. This is the reaction catalyzed by superoxide dismutases (SODs):

$$b.\ O_2^- + O_2^- + 2H^+ \rightarrow H_2O_2 + O_2$$

O_2^- is the conjugate base of a weak acid, the hydroperoxyl radical, which has a pK_a of ~4.8. Protonation of O_2^- greatly increases its ability to act as an oxidant. This can be illustrated by the effect of pH on the spontaneous dismutation reaction, which has a second order rate constant at 25° C of ~$10^5\ M^{-1}\ S^{-1}$ in acid solution. The rate constant increases to ~$10^8\ M^{-1}\ S^{-1}$ at pH = 4.8, and then decreases by a power of ten for each unit increase in pH above 4.8. In essence, the $HO_2\cdot + HO_2\cdot$ dismutation is fast and the $HO_2\cdot + O_2^- + H^+$ rate is even faster, while the $O_2^- + O_2^- + 2H^+$ dismutation does not occur in the absence of catalysis. Another indication of the enhancement of oxidative capacity of O_2^- by protonation is the rate of reaction with linoleic acid. Thus, O_2^- does not perceptibly oxidize linoleate but $HO_2\cdot$ does so at a rate of $1.2 \times 10^3\ M^{-1}\ S^{-1}$ (Bielski et al., 1983). Protonation is not the only way to enhance the oxidative propensity of O_2^-. Association with other cationic centers, such as Mn(II) (McPhail et al., 1976; Bielski and Chan, 1978) or V(V) (Liochev and Fridovich, 1986) exerts a similar effect.

In the biological milieu, O_2^- causes mischief primarily by acting as an oxidant. Among the low molecular compounds known to be oxidized by O_2^- are glutathione, ascorbate, tetrahydropterins, sulfite, leukoflavins, catecholamines and the enediolate tautomers of sugars and of sugar derivatives. Since O_2^- is a univalent oxidant its action can initiate self-propagating chain reactions which amplify the consequences of the initiating event. The net effect of such O_2^--initiated oxidations is depletion of cellular reductants with increased production of H_2O_2. SOD, by removing O_2^-, decreases such oxidations and spares cellular reductants while diminishing the production of H_2O_2.

Some investigators, who have observed deleterious consequences of overproduction of SOD, have argued that SOD increases the production of H_2O_2 (Yarom et al., 1988; Scott et al., 1987). Were O_2^- to be produced in isolation and were it quite stable in the absence of SOD, then this argument might be tenable. However, O_2^- in neutral aqueous solutions is not stable, but rather dismutes spontaneously with a rate constant of ~10^5 M^{-1} S^{-1}. Moreover, the interior of cells are rich in reductants whose oxidation can be initiated by O_2^-. Under these conditions SOD will have the effect of *decreasing* net production of H_2O_2. How then can one explain the negative consequences of the overproduction of SOD?

Overproduction of any one protein is probably accompanied by underproduction of other proteins, since the net protein content of the cell remains fairly constant. Moreover, overproduction of metalloproteins, such as the SODs, will tax the ability of the cell to supply the necessary prosthetic metal. Thus, by way of example, overproduction of Cu,ZnSOD limits the availability of Cu(II) needed for the maturation of other Cu(II)-containing enzymes. An additional explanation can be found in the orchestrated cellular response to O_2^- which, in *E. coli*, is referred to as the soxRS regulon (Demple and Amabile-Cuevas, 1991). This group of coordinately regulated proteins includes MnSOD, glucose-6-phosphate dehydrogenase, endonuclease IV, fumarase C and OmpF, among others which remain to be identified. All of these are important parts of the overall defensive response and overproduction of SOD has been shown to diminish induction of the other members of the soxRS regulon (Liochev and Fridovich, 1991, 1992).

V. O_2^-—MACROMOLECULAR TARGETS

O_2^- is able to oxidatively inactivate [4Fe-4S]-containing dehydratases, including aconitase, fumarase A+B, 6-phosphogluconate dehydratase, and dihydroxy acid dehydratase. Inactivation of these enzymes can explain some of the acute consequences of elevated intracellular levels of O_2^-. One such situation is the nutritional requirement, or auxotrophy, for branched chain amino acids which is imposed upon *E. coli* when O_2^- concentration is increased by hyperoxia (Brown and Yein, 1978), paraquat (Fee et al., 1981), or by a mutational deficit in SOD (Carlioz and Touati, 1986). Inactivation by O_2^- of the dihydroxy acid dehydratase, which catalyzes the penultimate step in the biosynthesis of branched chain amino acids, leaves the cells unable to grow unless the medium provides these amino acids.

The rate constants for both the inactivation of the *E. coli* aconitase by O_2^- and for its subsequent reactivation by cytosolic Fe(II) plus GSH have been estimated (Gardner and Fridovich, 1992). This allowed the steady state balance between active and inactive aconitase to serve as an index of intracellular O_2^- concentration under different conditions. That some fraction of cellular aconitase was inactive, presumably due to attack by O_2^- even in SOD-competent cells, indicates the magnitude of the threat posed by O_2^-. Further, what was shown for aconitase probably applies to the other [4Fe-4S]-containing dehydratases and perhaps to other classes of O_2^--sensitive enzymes that remain to be identified. That there are additional targets for O_2^- is shown by the fact that SOD-null mutants of *E. coli* exhibited O_2^--dependent auxotrophies not only for branched chain amino acids but for aromatic and for sulfur-containing amino acids as well (Carlioz and Touati, 1986).

VI. HO· FROM O_2^- + H_2O_2

We have already seen that O_2^-, *per se*, can damage both low molecular weight and high molecular weight components of living cells. It can also give rise to the indiscriminately reactive hydroxyl radical, HO·, and the way this happens in vivo is probably quite different from the way it occurs *in vitro*. *In vivo* O_2^- oxidizes the [4Fe-4S] clusters of dehydratases, as follows:

$$c.\ [2Fe(II)2Fe(III)\text{-}4S] + O_2^- + 2H^+ \rightarrow H_2O_2 + [Fe(II)3Fe(III)\text{-}4S]^+$$

Once the cluster has been oxidized in this way it becomes unstable and releases Fe(II), leaving behind the inactive [3Fe(III)-4S] cluster. The Fe(II) liberated from the oxidized cluster could then reduce H_2O_2 to HO^- + HO·, in what is called the Fenton reaction.

HO· is so reactive that were it produced in free solution within a cell it would react within the first collision or two, mostly with low molecular weight components such as sugars, amino acids, glutathione or metabolic intermediates, all of which are expendable targets. However, Fe(II) would not be likely to remain free, but would rather bind to polyanionic macromolecules such as DNA, or to polyanionic surfaces, such as the cell membrane. In that case the HO· generated by the Fenton reaction would be made adjacent to, and would selectively react with, these critical targets. If this scenario is correct then metal cations incapable of participating in the Fenton reaction might, by displacement of bound Fe(II), protect against the deleterious consequences of HO· production and iron-chelating agents should protect for a similar reason. Human fibroblasts were protected against H_2O_2 by iron-chelating agents (Mello Filho and Meneghini, 1984), as were bovine endothelial cells (Hiraishi et al., 1992), and Zn(II) protected the erythrocyte stroma against oxidation by an enzymatic source of H_2O_2 + O_2^- (Girotti et al., 1986).

In the *in vivo* setting O_2^- increases the availability of Fe(II) for the Fenton reaction by acting as an *oxidant* towards the [4Fe-4S] clusters of dehydratases. This can be contrasted with the *in vitro* case in which O_2^- acts as a *reductant* towards ferric complexes. Because of the almost inescapable contamination of phosphate buffer salts with iron, the production of HO· from O_2^- + H_2O_2 was first noted without appreciation of the catalytic role of Fe(III) (Beauchamp and Fridovich, 1970). Once the role of iron was clarified (McCord and Day, 1978) the biological significance of the process was questioned because of the abundance of intracellular reductants (Winterbourn, 1979, 1981). Thus how could O_2^- possibly compete with ascorbate and glutathione in bringing about the reduction of Fe(III)? It now appears that a broader question would have been more useful, i.e., how could O_2^- increase the supply of

Fe(II) that was available for reaction with H_2O_2? The seemingly paradoxical answer is that it does so *in vivo* by acting as an oxidant, not as a reductant, and by thus causing release of Fe(II) from [4Fe-4S] clusters.

VII. PARAQUAT: AN O_2-DEPENDENT MISCHIEF MAKER

This redox active bipyridylium herbicide has often been used to increase O_2^- production in a variety of cell types. It enters living cells, often by active uptake, and is then reduced univalently to the blue monocation radical. This radical is stable anaerobically, but it rapidly reduces O_2. Paraquat can thus serve to divert electron flow from normal metabolic pathways, which are useful to the cell, into a useless truncated pathway which produces O_2^-. In full accord with this view of paraquat action are its abilities to increase cyanide-resistant respiration (Hassan and Fridovich, 1977); exert an oxygen-dependent toxicity (Hassan and Fridovich, 1978); serve as an oxygen-dependent mutagen (Hassan and Moody, 1982); and induce the biosynthesis of MnSOD (Hassan and Fridovich, 1977) and of the other enzymes of the defensive soxRS regulon (Greenberg et al., 1990; Tsaneva and Weiss, 1990).

That the mischief caused by paraquat is at least partially attributable to O_2^- is made abundantly clear by the observation that SOD-null mutants of *E. coli* exhibit a markedly increased susceptibility towards the oxygen-dependent toxicity of paraquat (Carlioz and Touati, 1986). Moreover, this phenotypic deficit could be reversed by insertion of a functional gene coding for a foreign SOD (Natvig et al., 1987). Having established that O_2^- contributes to the toxicity of paraquat, we should consider whether there are also O_2^--independent mischiefs.

In order for paraquat to cause intracellular O_2^- production it must be reduced and this requires that there be paraquat reductases, or diaphorases. *E. coli* has been reported to contain a soluble NADPH: paraquat diaphorase (Hassan and Fridovich, 1979). Fractionation of extracts of *E. coli* by HPLC have revealed that there are several NADPH:paraquat diaphorases in *E. coli* and that one of them is flavodoxin reductase (Liochev and Fridovich, 1993). More recent work has revealed that there are also NADH: paraquat diaphorases. The multiplicity of paraquat diaphorases in *E. coli* explains why a search for paraquat-resistant mutants revealed strains with a deficit in paraquat uptake, but none with a defect in paraquat reduction (Kao and Hassan, 1985).

Diversion of electron flow from thioredoxin reductase to paraquat deprives the cell of its supply of reduced thioredoxin and that inconveniences the cell in ways which cannot be addressed by SOD. A similar consideration applies to each of the other paraquat diaphorases whose true identities and normal functions remain to be discovered. This explains why gross overproduction of SOD by genetic manipulations (Yarom et al., 1988; Scott et al., 1987) has usually failed to provide heightened resistance towards paraquat. Thus, the wild type level of SOD is sufficient to minimize toxic effects due to O_2^- production by paraquat, leaving in place the deficits caused by diversion of electron flow from normal pathways.

VIII. NITRIC OXIDE: ANOTHER RADICAL

There seems no end to the surprises which are turned up by careful study of living things. A recently uncovered example of such surprises is the biosynthesis of NO from L-arginine and the multiple functions that NO serves. NO is the endothelium-derived relaxing factor that regulates the contraction of the smooth muscle in the walls of blood vessels. It controls how much blood flows in different parts of the circulatory tree and also the pressure of that blood. NO is also a neurotransmitter and it is the cytocidal agent released from cytotoxic macrophages. Given that knowledge of NO in the biological setting is only 6 years old and is expanding rapidly, we can be certain that additional functions will come to light.

We will consider NO only so far as it bears upon oxyradicals. Both NO and O_2^- are free radicals and should be expected to react with each other at a diffusion-limited rate to yield peroxynitrite, and they do so (Huie and Padmaja, 1993). Peroxynitrite is a powerful oxidant and can also be converted to nitrate. Before the NO story had begun to unfold, surviving slices of rat liver were seen to convert ammonia to nitrate, and paraquat increased the yield of nitrate (Dull and Hotchkiss, 1984). Before the endothelium derived relaxing factor had been identified as NO, its lifetime was seen to be extended by SOD (Auch-Schwalk et al., 1989), as was its yield from the irradiation of nitrite (Matsunaga and Furchgott, 1989). All of this can now be explained as consequences of the rapid reaction of NO with O_2^-.

NO through its action as the relaxing factor exerts a hypotensive effect, and O_2^-, by converting NO to peroxynitrite and thus inactivating it, exerts a hypertensive effect. In that case SOD, by eliminating O_2^- and thereby extending the lifetime of NO, should have a hypotensive effect. SOD has indeed been seen to cause arteriolar relaxation (Langenstroer and Pieper, 1992) and injection of SOD has been reported to lower the blood pressure of spontaneously hypertensive rats (Nakazono et al., 1991). In the latter work an SOD with a heparan sulfate-binding domain was used so that it would attach itself to the endothelium. The engineered SOD used in these studies was very much like the extracellular SOD (ECSOD) produced endogenously, and modulating the interaction of O_2^- with NO is probably one of its functions. The biology of NO and the consequences of its reaction with O_2^- will be discussed more fully by Freeman and Beckman in a later chapter of this volume.

IX. SUPEROXIDE DISMUTASES: THE FIRST LINE OF DEFENSE

Even though O_2^- is intrinsically unstable in neutral aqueous solutions due to the spontaneous dismutation reaction, it can last a long time if its concentration is low.

In essence, the spontaneous dismutation requires that one O_2^- (or $HO_2^.$) find another to react with. The more dilute the solution the longer that will take. Let us calculate the time required for half of a given concentration of O_2^- to dismute, i.e. its first half life. The rate constant for the spontaneous dismutation at 25° C and at pH 7 is close to 1×10^5 M^{-1} S^{-1} and the rate at which O_2^- would be consumed would be equal to $[O_2^-]^2$ times 1×10^5 M^{-1} S^{-1}. When $[O_2^-]$ is 1 mM then the first half life is only 5 msec, but it increases to 5 seconds at 1 μM O_2^- and to 83 min at 1 nM O_2^-!

SODs profoundly shorten the half life of O_2^- and have their greatest effect when the concentration of O_2^- is low. Thus SOD functions by taking an electron from one O_2^- and then passing that electron to a second O_2^-. In this way the dismutation involves reactions of O_2^- with the enzyme rather than directly with another O_2^-, and the rate depends upon the first power of $[O_2^-]$ rather than upon its square. The rate constant for the reaction of O_2^- with SOD is approximately 3×10^9 M^{-1} S^{-1} and the concentration of SOD in many cells is close to 10^{-5}M. Under these conditions the rate of consumption of O_2^- would be equal to 3×10^9 M^{-1} S^{-1} times 1×10^{-5} M times $[O_2^-]$ and its half life would be 17 μsec, regardless of the initial concentration of the radical.

This strategy of using SOD to mediate the electron transfer from one O_2^- to the next, and thereby to shorten the lifetime of O_2^-, has been exploited to the full and SODs of several different types are abundant in aerobic organisms. In mammals we find three types of SODs which represent two distinct evolutionary histories and each of which is found in a different location. Thus in cellular cytosols we find a Cu,ZnSOD with a molecular weight of ~32,000 and composed of two identical subunits. In mitochondria we find a MnSOD whose molecular weight is ~92,000 and is a homotetramer. Finally, in the extracellular spaces there is a Cu,ZnSOD (ECSOD) that differs from the cytosolic enzyme in being tetrameric and glycosylated and in having a larger subunit size. It moreover exhibits an affinity for heparan sulfate and so binds to cell surfaces. There is sequence homology between the Cu,ZnSODs of the cytosol and of the extracellular space, which indicates an evolutionary relationship. In contrast, the mitochondrial MnSOD shows no sequence homology to the Cu,ZnSODs and this bespeaks an independent evolutionary origin.

Each of these mammalian SODs is an equally superb catalyst of the dismutation reaction, exhibiting rate constants close to the theoretical diffusion limit. Why then do we need three different SODs? The answer lies both in functional requirements and in evolutionary history. Thus an enzyme as small as the cytosolic Cu,ZnSOD would not last long in the extracellular spaces, because it would be filtered out by the kidneys. Hence the advantage of a larger, tetrameric ECSOD that cannot be filtered by the kidneys and clings to the heparan sulfate of the extracellular matrix.

The mitochondrial MnSOD requires another type of explanation. It may have properties that make it specially suited for the mitochondrial matrix, but we do not yet know what they are. There is, however, an explanation embedded in the evolutionary origin of mitochondria. There is thus abundant support for the view that mitochondria evolved from prokaryotic endosymbionts (Gray et al., 1989), and prokaryotes commonly contain MnSODs which exhibit extensive sequence homologies to the mitochondrial MnSOD. The mitochondrial SOD is thus a relic of the bacterial origin of these organelles. Many bacteria also contain a FeSOD that exhibits sequence homologies to the MnSODs, but no mitochondrial FeSOD has yet been found.

There is another deduction to be made concerning the multiple SODs found in mammals. Were O_2^- relatively unreactive and free to diffuse across organellar and plasma membranes then O_2^- made in one compartment could be dealt with by SOD in another locale. However, as we have already discussed, O_2^- is very reactive in the biological milieu and, being charged and hydrophilic, it will not readily cross biological membranes. It follows that O_2^- made in a particular locale, must be eliminated by SOD in that locale. There is experimental support for that view in that a deficiency of the mitochondrial MnSOD in yeast can be complemented by functional MnSOD genes, derived from tobacco (Bowler et al., 1989), or from maize (Zhu and Scandalios, 1992); provided that the sequence coding for the transit peptide, which ensures importation into the mitochondria, is intact.

X. THE CONSEQUENCES OF DOING WITHOUT

Early studies of the biological effects of O_2^- and of the defensive role of the SODs, were dependent upon manipulation of O_2^- production by changing pO_2, and by adding redox active viologens and quinones, and upon observing the natural induction of SODs and the consequences of such inductions (Fridovich, 1989). More recent and more convincing studies have utilized mutational elimination and restoration of SODs. SOD-null *E. coli* exhibited an oxygen-dependent inability to grow in minimal medium. They could grow in rich medium but growth was then slower aerobically than anaerobically. These deficits were complemented by a plasmid bearing an SOD gene (Carlioz and Touati, 1986). O_2-dependent spontaneous mutagenesis was also increased by a lack of SOD. This mutagenesis was exacerbated by the redox cycling quinone plumbagin, and was eliminated by restoration of SOD (Farr et al., 1986). The phenotypic deficits exhibited by SOD-null *E. coli* could be eliminated by expression of any functional SOD. Thus, a gene coding for the human Cu,ZnSOD was able to restore normal ability to grow in aerobic minimal medium, as well as the normal resistance towards paraquat and towards H_2O_2 (Natvig et al., 1987).

Similar studies have since been done with other organisms and in all cases lack of SOD imposes O_2^--dependent deficits which could be reversed by reinsertion of an SOD gene. It is clearly hard to do without SOD in an aerobic world. There are nevertheless aerotolerant organisms that manage without SOD. Among these is *Neisseria gonorrhoeae* (Norrod and Morse, 1979) which may compensate for its lack of SOD by accumulating almost 100 times more catalase than is found in *E. coli* (Hassett et al., 1990), or by the replacement of O_2^--sensitive enzymes by resistant analogues.

XI. *LACTOBACILLUS PLANTARUM* AND A FUNCTIONAL REPLACEMENT FOR SOD

One family of organisms, exemplified by *L. plantarum*, teaches us another lesson about the versatility of living things. It has dispensed with the need for SOD by having found a non-enzymic replacement for it. *L. plantarum* normally lives in fermenting plant materials, which are always rich in manganese, and when grown in defined medium does best when Mn(II) is provided at a level of several hundred micromolar. *L. plantarum* was found to lack a true SOD; but its extracts did contain a dialyzable, heat-stable, and EDTA-sensitive ability to catalyze the dismutation of O_2^-. This activity was traced to Mn(II), which these cells concentrated from the medium to an intracellular level of ~25 mM (Archibald and Fridovich, 1981a). Growth in a relatively Mn(II)-deficient medium imposed an intolerance towards oxygen, indicating that the Mn(II) was serving this organism as a functional replacement for SOD. Further support for this conclusion came from a survey of different species of *Lactobacilli*, some of which accumulated Mn(II), and some of which contained true SOD, but none of which did both (Archibald and Fridovich, 1981b).

XII. CATALASES: MEMBERS OF THE DEFENSIVE TEAM

When considered in isolation, the dismutation of O_2^- into $O_2 + H_2O_2$ hardly seems like a bargain. The H_2O_2 produced by the dismutation reaction is by no means benign and it can, by reacting with metals such as Fe(II), Cu(I), and V(IV), give rise to the terribly reactive HO·. Fortunately the SODs do not act alone. They are helped by the catalases which dismute H_2O_2 into $O_2 + H_2O$.

Mammalian catalase is a large enzyme of 240 kD. It is a homotetramer with ferriheme at its active sites. It is largely, but not exclusively, found in the subcellular organelles called peroxysomes. The advantage of such localization can best be understood in terms of the kinetic behavior of catalase, which exhibits a high K_m for H_2O_2. This means that catalase will act most efficiently in the decomposition of H_2O_2 when presented with this substrate at high concentration. The peroxysome is also the site of localization of many H_2O_2-producing enzymes and co-localization of catalase with these enzymes allows the catalase to act upon the H_2O_2 before it has been diluted by diffusion into the total volume of the cell.

Catalase is versatile and can use H_2O_2 to oxidize not only H_2O_2, but also a variety of small molecules, such as methanol, ethanol, formate, nitrite, and mercury. It cannot peroxidize larger molecules, presumably because the heme prosthetic is enclosed in a crevice which excludes them (Sichak and Dounce, 1986). The physiological importance of the limited peroxidatic activity of catalase remains unclear. This is particularly the case since there are dedicated peroxidases, such as the glutathione peroxidase, which use GSH to reduce not only hydrogen peroxide, but alkyl hydroperoxides as well. In spite of an enormous amount of study, there are aspects of catalase action which are still inexplicable. One recent example is the ability of catalase in the medium to prevent apoptosis of a mammalian cell line cultured in the absence of serum (Sandstrom and Buttke, 1993). Another is the stabilization of catalase by NADPH and the autoinactivation of catalase in glucose-6-phosphate-deficient, and therefore, NADPH-poor, human erythrocytes.

XIII. SUMMARY

We have briefly examined the basis of the oxidant stress facing all aerobic cells and have considered a few of the enzymatic defenses that have evolved to protect against this stress. There is much more, already uncovered, and vast areas that are not yet illuminated. We have not considered the important roles played by antioxidants, both hydrophilic and lipophilic, nor have we spoken of the enzymes dedicated to the repair and recycling of oxidatively-damaged macromolecules. We do now understand, albeit incompletely, that oxygen-derived free radicals contribute to numerous physiological and pathological processes, including senescence. Further exploration is warranted and the payoffs are apt to be generous. Certainly the Duc of Rochefoucald was correct in saying that "knowledge is the only way out of the cages of life," and we are busily pursuing that knowledge. There will be more!

ACKNOWLEDGMENTS

This work was supported by a grant from the Council for Tobacco Research-U.S.A., Inc. and a Johnson and Johnson Focused Giving Grant.

REFERENCES

Archibald, F. S., & Fridovich, I. (1981a). Manganese and defenses against oxygen toxicity in Lactobacillus plantarum. J. Bacteriol. 145, 442-451.

Archibald, F. S., & Fridovich, I. (1981b). Manganese, superoxide dismutase and oxygen tolerance in some lactic acid bacteria. J. Bacteriol. 146, 928-936.

Auch-Schwalk, W., Kautusic, Z., & Vanhoutte, P. (1989). Contractions to oxygen derived free radicals are augmented in aorta of the spontaneously hypertensive rat. Hypertension 13, 859-864.

Beauchamp, C., & Fridovich, I. (1970). A mechanism for the production of ethylene from methional: the generation of hydroxyl radical by xanthine oxidase. J. Biol. Chem. 245, 4641-4646.

Beauchamp, C., & Fridovich, I. (1971). Superoxide dismutase: Improved assays and an assay applicable to acrylamide gels. Anal. Biochem. 44, 276-287.

Bielski, B. H. J., Arudi, R. L., & Sutherland, M. W. (1983). A study of the reactivity of perhydroxyl radical/superoxide anion with unsaturated fatty acids. J. Biol. Chem. 258, 4759-4761.

Bielski, B. H. J., & Chan, P. C. (1978). Products of reaction of superoxide and hydroxyl radicals with the Mn(II) cation. J. Am. Chem. Soc. 100, 1920-1921.

Bowler, C., Alliotte, T., Van den Bulcke, M., Bauw, G., Vanderkerckhove, J., Van Montagu, M., & Inzé, D. (1989). A plant manganese superoxide dismutase is efficiently imported and correctly processed by yeast mitochondria. Proc. Nat. Acad. Sci. USA 86, 3237-3241.

Brown, O. R., & Yein, F. (1978). Dihydroxy acid dehydratase: the site of hyperbaric oxygen poisoning in branched chain amino acid biosynthesis. Biochem. Biophys. Res. Commun. 85, 1219-1224.

Carlioz, A., & Touati, D. (1986). Isolation of superoxide dismutase mutants in *Escherichia coli*: Is superoxide dismutase necessary for aerobic life? EMBO J. 5, 623-630.

Demple, B., & Amabile-Cuevas, C. F. (1991). Redox redux: the control of oxidative stress response. Cell 67, 837-839.

Dull, B. J., & Hotchkiss, J. H. (1984). Activated oxygen and mammalian nitrate biosynthesis. Carcinogenesis 5, 1161-1164.

Elstner, E. F., & Heupel, A. (1976). Inhibition of nitrite formation from hydroxylammonium chloride: a simple assay for superoxide dismutase. Anal. Biochem. 70, 616-620.

Farr, S. B., D'Ari, R., & Touati, D. (1986). Oxygen-dependent mutagenesis in *Escherichia coli* lacking superoxide dismutase. Proc. Nat. Acad. Sci. USA 83, 8268-8272.

Fee, J. A., Lees, A. C., Bloch, P. L., Gilliland, P. L., & Brown, O. R. (1981). Oxygen stasis of bacterial growth: Analogy between the stases of *Escherichia coli* by hyperbaric oxygen and by aerobic paraquat. Biochem. Int. 1, 304-311.

Fridovich, I. (1989). Superoxide dismutases. An adaptation to a paramagnetic gas. J. Biol. Chem. 264, 7761-7764.

Gardner, P. R., & Fridovich, I. (1992). Inactivation-reactivation of aconitase in *Escherichia coli*. A sensitive measure of superoxide radical. J. Biol. Chem. 267, 8757-8763.

Girotti, A. W., Thomas, J. P., & Jordan, J. E. (1986). Inhibitory effect of zinc(II) on free radical lipid peroxidation in erythrocyte membranes. J. Free Rad. Biol. Med. 1, 395-401.

Gray, M. W., Cedergren, R., Abel, Y., & Sankoff, D. (1989). On the evolutionary origin of the plant mitochondrion and its genome. Proc. Nat. Acad. Sci. USA 86, 2267-2271.

Greenberg, J. T., Monach, P., Chou, J. H., Josephy, P. D., & Demple, B. (1990). Positive control of a global antioxidant defense regulon activated by superoxide-generating agents in *Escherichia coli*. Proc. Nat. Acad. Sci. USA 87, 6181-6185.

Hassan, H. M., & Fridovich, I. (1977). Regulation of the biosynthesis of superoxide dismutase in *Escherichia coli*. J. Biol. Chem. 252, 7667-7672.

Hassan, H. M., & Fridovich, I. (1978). Superoxide radical and the oxygen-enhancement of the toxicity of paraquat in *Escherichia coli*. J. Biol. Chem. 253, 8143-8148.

Hassan, H. M., & Fridovich, I. (1979). Mechanism of production of extracellular superoxide radical. J. Biol. Chem. 254, 10846-10852.

Hassan, H. M., & Moody, C. S. (1982). Superoxide dismutase protects against paraquat-mediated dioxygen toxicity and mutagenicity: Studies in *Salmonella typhimurium*. Can. J. Physiol. Pharmacol. 60, 1367-1373.

Hassett, D. J., Charniga, L., & Cohen, M. S. (1990). *recA* and catalase in H_2O_2-mediated toxicity in *Neisseria gonorrhoeae*. J. Bacteriol. 172, 7293-7296.

Heikkila, R. E., & Cabbat, F. S. (1976). A sensitive assay for superoxide dismutase based on the autoxidation of 6-hydroxydopamine. Anal. Biochem. 75, 356-362.

Hiraishi, H., Terano, A., Razandi, M., Sugimoto, T., Harada, T., & Ivey, K. J. (1992). Role of superoxide dismutase against reactive oxygen metabolite injury in cultured bovine aortic endothelial cells. J. Biol. Chem. 267, 14812-14817.

Huie, R. E. & Padmaja, S. (1993). The reaction of NO with superoxide. Free Radical Res. Commun. 18, 195-199.

Imlay, J. A., & Fridovich, I. (1991). Assay of metabolic superoxide production in *Escherichia coli*. J. Biol. Chem. 266, 6957-6965.

Kao, S. M., & Hassan. H. M. (1985). Biochemical characterization of a paraquat tolerant mutant of *Escherichia coli*. J. Biol. Chem. 260, 10478-10481.

Langenstroer, P., & Pieper, G. M. (1992). Regulation of spontaneous EDRF release in diabetic rat aorta by oxygen free radical. Am. J. Physiol. 263, 4257-4265.

Liochev, S., & Fridovich, I. (1986). The vanadate-stimulated oxidation of NAD(P)H by biomembranes is a superoxide-initiated free radical chain reaction. Arch. Biochem. Biophys. 250, 139-145.

Liochev, S. I., & Fridovich, I. (1991). Effects of overproduction of superoxide dismutase on the toxicity of paraquat toward *Escherichia coli*. J. Biol. Chem. 366, 8747-8750.

Liochev, S. I., & Fridovich, I. (1992). Effects of overproduction of superoxide dismutases in *Escherichia coli* on inhibition of growth and on induction of glucose-6-phosphate dehydrogenase by paraquat. Arch. Biochem. Biophys. 294, 138-143.

Liochev, S. I., & Fridovich, I. (1994). Paraquat diaphorases in *Escherichia coli*. Free Rad. Biol. Med.16, 555-559.

Marklund, S., & Marklund, G. (1974). Involvement of the superoxide radical in the autoxidation of pyrogallol and a convenient assay for superoxide dismutase. Eur. J. Biochem. 47, 469-474.

Matsunaga, K., & Furchgott, R. F. (1989). Interactions of light and sodium nitrite in producing relaxation of rabbit aorta. J. Pharm. Exp. Ther. 248, 687-695.

McCord, J. M., & Fridovich, I. (1969). Superoxide dismutase. An enzymic function for erythrocuprein (hemocuprein). J. Biol. Chem. 244, 6049-6055.

McCord, J. M., & Day, E. D., Jr. (1978). Superoxide-dependent production of hydroxyl radical catalyzed by iron-EDTA complex. FEBS Lett. 86, 139-142.

McPhail, L. C., DeChatelet, L. R., & Shirley, P. S. (1976). Further characterization of the NADPH oxidase activity of human polymorphonuclear leukocytes. J. Clin. Invest. 58, 775-780.

Mello Filho, A. C., & Meneghini, R. (1984). *In vivo* formation of single strand breaks in DNA by hydrogen peroxide is mediated by the Haber Weiss reaction. Biochim. Biophys. Acta 781, 56-63.

Nakazono, K., Watanabe, N., Matsuno, K., Sasaki, J., Sato, T., & Inoue, M. (1991). Does superoxide underlie the pathogenesis of hypertension? Proc. Nat. Acad. Sci. USA 88, 10045-10048.

Natvig, D. D., Imlay, K., Touati, D., & Hallewell, R. A. (1987). Human copper-zinc superoxide dismutase complements superoxide dismutase deficient *Escherichia coli* mutants. J. Biol. Chem. 262, 14697-14701.

Norrod, P., & Morse, S. A. (1979). Absence of superoxide dismutase in some strains of *Neisseria gonorrhoeae*. Biochem. Biophys. Res. Commun. 90, 1287-1294.

Sandstrom, P. A., & Buttke, T. M. (1993). Autocrine production of extracellular catalase prevents apoptosis of human CEM-T-cell line in serum free medium. Proc. Nat. Acad. Sci. USA 90, 4708-4712.

Scott, M. D., Meshnick, S. R., & Eaton, J. W. (1987). Superoxide dismutase-rich bacteria. Paradoxical increase in oxidant toxicity. J. Biol. Chem. 262, 3640-3645.

Scott, M. D., Wagner, T. C., & Chiu, D. T. (1993). Decreased catalase activity is the underlying mechanism of oxidant susceptibility in glucose-6-phosphate-deficient erythrocytes. Biochim. Biophys. Acta 1181, 163-168.

Sichak, S. P., & Dounce, A. L. (1986). Analysis of the peroxidatic mode of action of catalase. Arch. Biochem. Biophys. 249, 286-295.

Tsaneva, I. R., & Weiss, B. (1990). SoxR, a locus governing a superoxide response regulon in *Escherichia coli* K12. J. Bacteriol. 172, 4197-4205.

Winterbourn, C. C. (1979). Comparison of superoxide with other reducing agents in the production of $HO^{\cdot}$. Biochem. J. 182, 625-628.

Winterbourn, C. C. (1981). Hydroxyl radical production in body fluids. Roles of metal ions, ascorbate, and superoxide. Biochem. J. 198, 125-131.

Yarom, R., Sapoznikov, D., Havivi, Y., Avraham, K. B., Schickler, M., & Groner, Y. (1988). Premature aging changes in neuromuscular junctions of transgenic mice with an extra human Cu,ZnSOD gene: A model for tongue pathology in Down's syndrome. J. Neurol. Sci. 88, 41-53.

Zhu, D., & Scandalios, J. G. (1992). Expression of the maize MnSOD (sod3) gene in MnSOD-deficient yeast rescues the mutant yeast under oxidative stress. Genetics 131, 803-809.

REGULATION OF GENE EXPRESSION BY OXIDATIVE STRESS

Klaus Schulze-Osthoff and Patrick A. Baeuerle

Advances in Molecular and Cellular Biology
Volume 25, pages 15-44.

ISBN: 0-7623-0379-4

I. INTRODUCTION

Bacteria and higher eukaryotes have developed elaborate mechanisms to rapidly respond to changes in their environment. In many cases, coupling of a ligand to a receptor triggers a signal that is transduced into the cytoplasm and stimulates the synthesis of new proteins by the induction of certain genes in the cell nucleus. Important elicitors of gene induction are diverse signals constituting different forms of cellular stresses. In particular, reactive oxygen intermediates (ROIs) are widely used mediators of stress responses. Many adverse conditions including heat shock, ionizing and UV irradiation, viral and bacterial infections, and environmental pollutants are known to cause oxidative stress. Several proteins have been identified that are newly synthesized in response to oxidative stress and may either confer protection against subsequent adversities, repair ROI-mediated damage of cellular components, or alert neighboring cells. The synthesis of new proteins is in most cases regulated at the transcriptional level by the activation of factors binding to regulatory DNA sequences of target genes. Research of the past few years has identified several transcription factors that are activated by changes of the intracellular ROI levels. In this review, we will describe and compare mechanisms that have evolved to translate ROI-mediated signals into transcriptional activation. We will especially focus on the redox regulation of NF-κB and AP-1, two important eukaryotic transcription factors activated by redox-dependent processes. The oxidative stress response of bacteria will also be covered.

II. GENERAL MECHANISMS OF GENE TRANSCRIPTION

In both bacteria and higher organisms gene expression is controlled by sequence-specific promoter elements which allow oriented binding of DNA-dependent RNA polymerases. Eukaryotic genes are mostly transcribed by RNA polymerase II (Pol II). The accurate initiation of transcription by Pol II is regulated by controlled interactions between two different types of DNA-binding proteins. The first are a group of ubiquitous, general factors required for recognition of the transcriptional start site, the assembly of the preinitiation complex and the subsequent recruitment of the polymerase. These basal transcription factors, including TFII-A, -B, -D, -E, -H, and -I, interact with Pol II in the initiation of transcription. Formation of the preinitiation complex is coordinated by binding of the TATA box-binding protein (TBP) and associating factors (TAFs) to the TATA box element (reviewed in Roeder, 1991). A second class of transcription factors regulate either tissue-specific or inducible gene expression. These activators recognize site-specific sequences in the upstream promotor region and can dramatically increase or decrease the rate of transcription. The interaction between basal and specific transcription factors is poorly understood. Some recent reports suggest a direct physical association between specific and basal factors (Zhu et al., 1991).

Various mechanisms have evolved to induce gene transcription in response to extracellular signals. One widespread mechanism is *de novo* synthesis of a transcription factor. *De novo* synthesis, however, is not a direct event and requires one or more "primary" transcription factors to be activated. For immediate-early gene expression various posttranslational mechanisms have evolved. In some cases, inhibitory proteins play a role in controlling activation of transcription. One important example is the transcription factor NF-κB. NF-κB resides in the cytoplasm of uninduced cells as an inactive complex with its inhibitory subunit IκB (Baeuerle and Baltimore, 1988). Exposure to many stimuli rapidly activates NF-κB by dissociation of IκB. This allows translocation of the factor into the nucleus and subsequent DNA binding. A further mechanism of activating transcription factors is binding of accessory proteins. A paradigm is the serum response factor (SRF) which is only active after association on DNA with the ternary complex factor (TCF) (reviewed in Treisman, 1992; Janknecht and Nordheim, 1993). Perhaps the most common mechanism of controlling the action of preexisting transcription factors is covalent modification of their amino acid residues. Frequently, this involves phosphorylation or dephosphorylation of serine or threonine residues in the DNA binding or transactivation domain of the factor (reviewed by Hunter and Karin, 1992). Examples are activation of the cAMP response binding protein (CREB) by phosphorylation by protein kinase A in response to increased cAMP levels (Lee et al., 1990), and of c-Jun by dephosphorylation in response to activation of protein kinase C (Boyle et al., 1991). Other posttranslational modifications with regulatory potential are reduction-oxidation (redox) reactions. Several transcription factors contain a conserved cysteine residue in their DNA binding region (Xanthoudakis et al., 1992; Kumar et al., 1992). In cell-free systems, oxidation of such cysteine residues has been shown to decrease DNA binding. In the following, we will show that oxidative stress-controlled gene expression involves the regulation of transcription factors at several distinct levels.

III. CONTROL OF OXIDATIVE STRESS RESPONSES IN BACTERIA

The induction of genes in response to adverse environmental conditions has intensively been studied for several years in bacteria. In prokaryotes, functionally related genes are organized in cassettes, called regulons, which allow a coordinate control of gene expression. Several regulons specifically respond to environmental stress signals. For instance, DNA damage leads in *Escherichia coli* to gene activation of the *rec A* regulon, which controls the expression of some 20 genes, known as the SOS response. These genes induce a number of phenotypic alterations and enhance the capacity for DNA repair. Other systems such as the *mer* regulon have evolved to confer resistance to mercury toxicity. Another regulon serves to counteract the detrimental consequences of oxidative stress (reviewed in Demple, 1991; Storz et al., 1990a). When bacteria are treated with hydrogen peroxide, the synthesis of at least 30 proteins is induced. As a consequence, the cells become resistant to a subsequent treatment with hydrogen peroxide that would otherwise be lethal. In *E. coli* and *Salmonella typhimurium*, two regulons have been identified that are induced by hydrogen peroxide and superoxide anion, respectively. The *oxyR* regulon mediates H_2O_2-inducible expression of eight proteins, including the genes for catalase (*katG*) and NADPH-dependent alkyl hydroperoxidase (*ahpFC*). Activation is controlled by the transcription factor OxyR which binds to a rather large sequence of about 45 base pairs in the *oxyR* regulon. Exposure to even mild oxidizing conditions activates OxyR and induces transcription of the genes by σ^{70} RNA polymerase (Storz et al., 1990b). The transcriptional activation but not the DNA binding is reversed upon exposure to highly reducing conditions. Although OxyR is specifically activated upon exposure to H_2O_2, the mechanism by which the protein senses oxidizing conditions remains unknown. Mutational analyses revealed that reduction of a critical disulfide bridge is not involved. It has been speculated that a loosely bound metal ion or cysteine oxidation products such as sulfenic or sulfinic acid may mediate oxidant-induced OxyR activation (Storz et al., 1990a; Demple and Amabile-Cuevas, 1991).

A further oxidative stress-inducible system of bacteria is the *soxRS* regulon (Amabile-Cuevas and Demple, 1991; Wu and Weiss, 1991). Activation is specifically induced by superoxide anions or $O_2^{\cdot-}$-producing agents, such as paraquat or quinones. The *soxRS* system provides protection against oxidative stress by increasing the synthesis of manganese superoxide dismutase (*sodA*), the DNA repair enzyme endonuclease IV, glucose-6-phosphate dehydrogenase and at least six other proteins of unknown function. Induction of the *soxRS* regulon involves the redox-mediated activation of preexisting SoxR protein as the initial step in the signaling cascade. This might occur via oxidation of an iron-sulfur center leading to a change in protein conformation (Hidalgo and Demple, 1994). Activated SoxR then induces the *de novo* synthesis of a second transcription factor, SoxS, which finally activates transcription of the defense genes. This system represents an example where an adaptive response requires both a preexisting (primary) and a newly synthesized (secondary) transcription factor. It further illustrates that bacteria can discriminate between oxidative stress reactions induced by hydrogen peroxide and superoxide anions.

IV. REDOX REGULATION IN EUKARYOTES

Gene expression in mammalian cells is more complex than in bacteria and little is still known about the underlying mechanisms. The response to ROI involves the activation of numerous functionally unrelated genes associated with signal transduction, proliferation, and immunologic defense reactions. By analogy to bacteria, it is assumed that the mammalian response to ROI serves a protective function. However, most of the genes induced by oxidative stress can be equally activated by more physiological signals, such as growth factors or cytokines. This indicates that both groups of signals converge into the same pathway by sharing signaling molecules. Experimental evidence indeed suggests that this is the case. Physiological as well as ROI-triggered signals activate NF-κB and AP-1, two important and widely used transcription factors. As will be described below, the overlapping effects of both signals may also be explained by the fact that some physiological inducers seem to utilize ROIs as intracellular signaling molecules. In eukaryotes, ROIs may therefore act as second messenger molecules that integrate the diversity of gene-inducing signals into a common genetic response. Hence, ROI-induced gene expression might not be restricted to adverse environmental conditions, but has a more widespread and principal role in cellular metabolism.

V. NF-κB

NF-κB is a transcription factor that plays a crucial role in the regulation of numerous genes involved in cellular defense mechanisms in responses to immune and inflammatory processes (reviewed in Baeuerle, 1991; Grilli et al., 1993; Liou and Baltimore, 1993; Baeuerle and Henkel, 1994). A great variety of stimuli can activate NF-κB and initiate transcription of NF-κB-dependent genes. In the following, we will describe that the effect of perhaps all activating stimuli seems to rely on the induction of prooxidant conditions.

A. Structural and Functional Characteristics of NF-κB

NF-κB was first identified as a nuclear factor of mature B cells that specifically interacts with a decameric enhancer element (5'-GGGACTTTCC-3') of the immunoglobulin κ light chain gene (Sen and Baltimore, 1986). Very frequently, NF-κB is composed of the p50 and p65 subunits but at least three other subunits have been reported to participate in dimer formation. In addition to p50 and p65 (now also called RelA), p52, c-Rel and RelB have been identified (reviewed in Blank et al., 1992; Nolan and Baltimore, 1992; Liou and Baltimore, 1993). All share a homologous domain of about 300 amino acids which is required for dimerization, DNA binding, nuclear targeting and interaction with regulatory proteins. While p50 and p52 may be solely required for DNA binding, p65, c-Rel and

RelB have in addition transactivating activity. It has been shown that combinations of interactions between the different NF-κB subunits give rise to dimers with distinguishable DNA sequence and transcriptional specificity. In T cells and some other cell types, a transcriptionally inactive p50 homodimer can be found that occupies κB sites for active complexes (Kang et al., 1992). This negative regulatory effect of p50 homodimers was suggested to play a physiological role during T cell activation.

B. The Activation of NF-κB

A characteristic of NF-κB activation is that it does not require new protein synthesis. In non-stimulated cells, NF-κB is bound to a family of inhibitory proteins, called IκBs, which retain the complex in the cytoplasm and prevent DNA binding (Baeuerle & Baltimore, 1988; Schmitz et al., 1991; Beg and Baldwin, 1993). Activation of NF-κB involves the dissociation of IκB from the NF-κB complex. The best characterized inhibitor is IκB-α which prevents binding of DNA to p65, c-Rel, and RelB. All IκB proteins contain multiple, closely adjacent copies of a characteristic repeat structure of 30 amino acids, called SWI6/ankyrin repeats (Blank et al., 1992). Mutational analysis has revealed that these ankyrin repeats are necessary but not fully sufficient to prevent DNA binding (Inoue et al., 1992). It remains to be established how association of NF-κB subunits with different IκB proteins affects the mechanism of activation. Recent experimental evidence suggests that IκB-α is apparently not restricted to the cytoplasm, but may also occur within the nucleus (Zabel et al., 1993). Since IκB decreases the half-life of the NF-κB-DNA complex (Zabel and Baeuerle, 1990), and NF-κB upregulates transcription of the IκB-α gene (Brown et al., 1993; Sun et al., 1993), IκB-α may have an additional role as a negative regulator of NF-κB-dependent transcription in the nucleus.

Extracellular stimulation with different agents activates NF-κB within minutes by the release of IκB. This allows migration of the transcription factor into the nucleus and DNA binding. The release of IκB, which is the key event of NF-κB activation, is under intensive investigation. In cell-free systems, it has been shown that different protein kinases, such as protein kinase C (PKC) or haem-regulated eIF2 kinase, can activate the transcription factor by phosphorylation of IκB (Shirakawa and Mizel, 1989; Ghosh and Baltimore, 1990). However, in the case of the NF-κB inducers tumor necrosis factor (TNF) and interleukin-1 (IL-1), neither depletion nor inhibition of PKC suppresses NF-κB activation in intact cells (Meichle et al., 1990; Bomsztyk et al., 1991). Experiments using intacts cells suggest an involvement of Raf-1 kinase and the protein kinase C isoform ζ (Finco and Baldwin, 1993; Devary et al., 1993; Dominguez et al., 1993). Several studies have described that activation of NF-κB in response to many stimuli is paralleled by the degradation of IκB, suggesting that a proteolytic process is critically involved in the activation (Henkel et al., 1993; Sun et al., 1993; Brown et al., 1993). In intact cells, phosphorylation events could either target IκB for degradation or directly activate a specific protease (Figure 1).

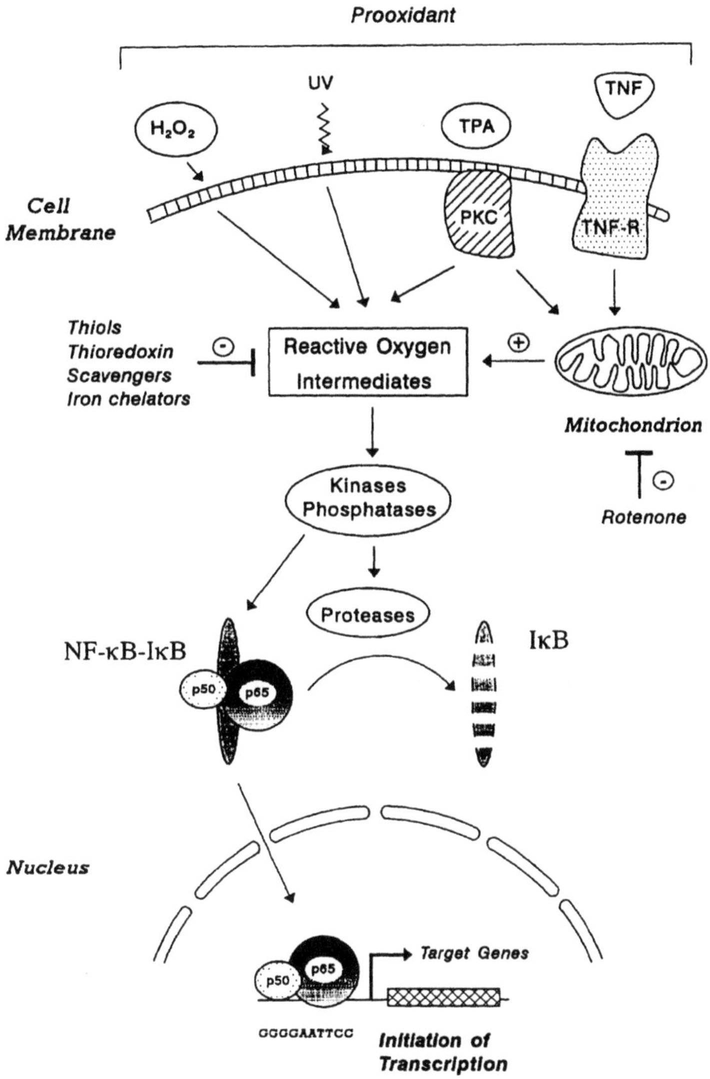

Figure 1. Redox regulation of NF-κB activation in response to hydrogen peroxide, UV irradiation, TPA and TNF: All reagents presumably activate NF-κB by the formation of reactive oxygen intermediates, because the activating effect is blocked by antioxidants. Intracellular ROI formation triggered by TNF is mainly originating during mitochondrial respiration. The key step of NF-κB activation involves the dissociation of IκB allowing the factor to translocate into the nucleus and to activate expression of target genes. NF-κB activation coincides with the proteolytic degradation of IκB and can be inhibited by certain protease inhibitors. Intermediate steps involved in the activation pathway may further include redox-mediated activation of protein kinases or phosphatases.

C. The Activation of NF-κB by Reactive Oxygen Intermediates

Potent stimuli activating NF-κB in intact cells are the cytokines TNF and IL-1, phorbol esters, lipopolysaccharides, double-stranded RNA, protein synthesis inhibitors, UV and ionizing irradiation and several viral transactivator proteins. Although many of these activators induce a similar pattern of genes as phorbol esters, protein kinase C does not seem to be involved in each of the signaling pathways (reviewed in Baeuerle and Henkel, 1994). A common feature of all inducers, however, is that they are pathogenic or at least proinflammatory stimuli. Interestingly, many, if not all NF-κB-inducing agents lead to a rise in the intracellullar concentrations of ROIs. In particular, TNF has been reported to induce oxidative stress in several cell types, which is thought to contribute to its cytotoxic and proinflammatory effects (Zimmerman et al., 1989; Yamauchi et al., 1989; Schulze-Osthoff et al., 1992). Since TNF is a potent inducer of NF-κB activation, we analyzed whether oxidative stress induced by exposure of cells to hydrogen peroxide can activate NF-κB. In the human T cell line Jurkat, 30 to 150 μM H_2O_2 indeed rapidly induced NF-κB DNA binding and transactivation activity (Schreck et al., 1991; 1992a). The activation of NF-κB in response to hydrogen peroxide was not restricted to lymphoid cells. For instance, in the cervix carcinoma line HeLa, 250 μM H_2O_2 strongly induced NF-κB activation (Meyer et al., 1993). Another reagent producing peroxides and potently stimulating NF-κB activation was butylperoxide. In contrast, several quinone compounds such as paraquat, menadione, doxorubicin or mitomycin, which continuously produce superoxide anions by redox-cycling, failed to induce NF-κB activation (Schreck et al., 1992b). In addition, the endoperoxide 3,3'-(1,4-naphthylidene)dipropionate ($NDPO_2$), a substance that releases singlet oxygen, did not cause NF-κB DNA binding. Nitric oxide (NO·), which has a well established signaling role in several biological processes (Ignaro, 1990), has recently been described to induce NF-κB in peripheral blood lymphocytes (Lander et al., 1993). However, our own experiments revealed that neither NO·-generating agents, such as nitroprusside, had an activating effect nor that inhibitors of NO· synthase were capable of suppressing TPA or TNF-induced NF-κB activation (Schreck et al., 1992b). Moreover, several other toxic conditions such as exposure to heavy metals (cadmium, copper salts) or heat shock did not lead to the activation of NF-κB. A recent report, however, showed that the transition metal nickel can induce NF-κB activation as well as expression of the NF-κB-dependent genes interleukin-6 and ELAM-1 in endothelial cells (Goebeler et al., 1995). Intriguingly, nickel ions play an important role in causing allergy and contact dermatitis.

D. The Effects of Antioxidants on NF-κB Activation

Further evidence for the idea that NF-κB activation relies on the intracellular generation of ROIs was provided by the effect of antioxidants. It was found that the thiol compounds N-acetyl-L-cysteine (NAC) or 2-mercaptoethanol and, more potently, pyrrolidine dithiocarbamate (PDTC) and related derivatives suppressed NF-κB activation induced by TNF or phorbol esters (Mihm et al., 1991; Schreck et al., 1991; 1992a). Up to now, all inducers of NF-κB, including interleukin-1, LPS, cycloheximide and virally encoded transactivator proteins from hepatitis B virus and human T cell leukemia virus I were blocked by these antioxidants (Schreck et al., 1992a; 1992b; Meyer et al., 1992). These findings indicate that formation of ROIs is indeed a common denominator of all NF-κB-inducing agents. While thiol-containing agents might primarily inhibit NF-κB activation by a radical scavenging effect, PDTC (and also the metal chelator o-phenanthroline) might, in addition, deplete free iron and copper ions that are necessary for the Fenton and Haber-Weiss reaction (Halliwell and Gutteridge, 1990). The suppressing effect of metal chelators further suggests that hydrogen peroxide has to be converted into other intermediates and that superoxide anion or, even more likely, hydroxyl radical or peroxides might be the NF-κB-inducing elicitors. In addition, lipid peroxidation may be important for NF-κB activation. Lipid-soluble chain-breaking antioxidants, such as butylated hydroxyanisole (BHA), nordihydroguaiaretic acid (NDGA) and vitamin E, are potent inhibitors of NF-κB activation (Israël et al., 1992; Schulze-Osthoff et al., 1993; Suzuki and Packer, 1993). On the other hand, hydrophilic derivatives of vitamine E inhibit NF-κB by two orders of magnitude better than unmodified lipophilic vitamin E (Suzuki and Packer, 1993).

NF-κB is probably not activated by ROIs in a direct manner, since neither H_2O_2 nor the xanthine oxidase system dissociates IκB from purified NF-κB-IκB complex under cell-free conditions. More conceivable are intermediate steps which can sense and are modulated by redox reactions. In this respect, redox regulatory molecules such as glutathione or thioredoxin may determine NF-κB activation. As will be discussed later, these molecules may control redox-sensitive signal transducers, such as certain kinases or proteases, which are thought to be involved in NF-κB induction.

VI. ACTIVATOR PROTEIN-1

Activator protein-1 (AP-1) is another important transcription factor which is regulated by redox-dependent processes. In contrast to NF-κB, AP-1 is exclusively localized in the nucleus. AP-1 has mainly been implicated in various processes associated with cell growth and differentiation events (reviewed in Angel and Karin, 1991; Karin and Smeal, 1992). Among the genes known to contain *cis*-regulatory AP-1 elements are those encoding the genes for metallothionein, collagenase, stromelysin, transforming growth factor β1 and interleukin-2. In the past years, AP-1 has attracted considerable interest, because it is composed of the *c-jun* and *c-fos* proto-oncogene product which have an oncogenic potential. Fos and Jun proteins form homo- (Jun/Jun) or heterodimeric (Jun/Fos) complexes that bind to a transcriptional control element, called the AP-1 site or TPA-responsive element

(TRE). Protein interaction occurs through a coiled-coil structure involving basic leucine zipper domains. Several other members of the family, including Jun B, Jun D, and the Fos-related activators Fra-1 and Fra-2 have been identified (reviewed in Angel and Karin, 1991). Heterodimer formation among the different members of the Fos/Jun family as well as interactions with ATF/CREB members generates a diverse array of protein complexes with overlapping DNA-binding specificities but distinct functional properties.

A. Activation of AP-1 by Prooxidants

The activity of AP-1 is mainly controlled by new protein synthesis. Expression of c-Jun and c-Fos synthesis is rapidly and transiently induced by a variety of different extracellular stimuli. These include mitogenic signals, phorbol esters, several differentiation-inducing stimuli and membrane depolarization. Most important in our context are various adverse conditions and stress signals, such as oxygen radicals, UV irradiation, exposure to heavy metals or DNA-alkylating agents. Intensive research over the past years has revealed that the regulation of AP-1 activity is highly complex and may differ in response to distinct signals. The mechanisms that regulate assembly and functional specificity of different AP-1 complexes are rather poorly understood. Differential expression of family members, interactions with unrelated transcription factors, conformational alterations and altered DNA binding specificities of the heterodimers are all likely to play a role.

HeLa cells, in which AP-1 activation has been most intensively studied, contain low amounts of primarily Jun homodimers in their uninduced state. Exposure to UV-light and to a lesser extent to hydrogen peroxide leads to a rapid enhancement of AP-1 DNA binding within a few minutes (Büscher et al., 1988; Nose et al., 1991, Devary et al., 1991). This immediate induction is not inhibited by cycloheximide and therefore independent of new protein synthesis. The prooxidant-induced activation primarily involves posttranslational modifications that are carried out by complex changes in the phosphorylation pattern of c-Jun. The phosphorylation in the DNA-binding domain is decreased resulting in enhanced DNA-binding activity, while phosphorylation of serine-63 and serine-73, located in the transactivation domain, is enhanced (Boyle et al., 1991; Binétruy et al., 1991). Superimposed to these posttranslational modifications is the *de novo* synthesis of AP-1 subunits, which may prolong the expression of target genes. Transcription of both *c-jun* and *c-fos* is rapidly induced within 30 min by exposure to hydrogen peroxide and other stimuli (Stein et al., 1989; Devary et al., 1991; Nose et al., 1991). The induction of *c-jun* transcription is thought to be mediated by an upstream AP-1 element. Hence, the gene is positively controlled by autoregulation through the initial activation of preexisting AP-1 or *c-jun* homodimers (Angel et al., 1988). The *c-fos* gene is not autoregulated but controlled by the serum response element (SRE) and various other elements responsive to cAMP, Ca^{2+} and mitogens (Angel and Karin, 1991; Treisman, 1992). The responsiveness of *c-fos* transcription to TPA, UV, hydrogen peroxide and antioxidants has been found to be conferred by the SRE (Stein et al., 1989; Shibanuma et al., 1990; J. Müller et al., 1997).

Although a multitude of different stimuli activates AP-1 and may lead to a similar pattern of induced genes, the used signaling pathway may be different among various inducers. For instance, TPA-treated cells are refractory to restimulation by the same stimulus, but remain fully responsive to UV irradiation and vice versa (Büscher et al., 1988; Devary et al., 1991). Comparison of the activation of AP-1 and NF-κB in response to TPA provides an example that one extracellular signal can obviously utilize different signal transduction pathways. As mentioned above, NF-κB activation by TPA is abolished by antioxidants but can be potentiated by hydrogen peroxide (Meyer et al., 1993). This indicates that NF-κB activation by TPA follows an ROI-dependent pathway. In contrast, H_2O_2 only weakly triggers AP-1 activation and even suppresses activation induced by TPA. This implies that, although TPA induces a prooxidant state in cells, an ROI-dependent pathway is only used for the activation of NF-κB but not for AP-1 activation.

B. Antioxidant-Induced AP-1 Activation

Contrary to the observations that *c-jun* and *c-fos* genes are strongly induced by prooxidant conditions, we were very surprised to find a strong induction of AP-1 DNA binding and transactivation under antioxidant conditions as well. Several structurally unrelated antioxidants, such as BHA, PDTC, 2-mercaptoethanol or dithiothreitol, which block NF-κB activation, strongly induced AP-1 activation (Meyer et al., 1993; Schenk et al., 1994). HeLa cells where AP-1 shows a very weak response to H_2O_2 treatment, revealed a strong increase of AP-1 activity by these antioxidants. The protein oxidoreductase thioredoxin, a potent physiological antioxidant, was also effective. Both transient overexpression or addition of recombinant thioredoxin to the culture medium resulted in enhanced AP-1 DNA-binding and transactivating activity (Meyer et al., 1993; Schenk et al., 1994; Amstad et al., 1991). In contrast to activation by TPA, antioxidant-mediated induction of AP-1 did not involve protein kinase C and was entirely dependent on *de novo* protein synthesis. Following treatment of cells with PDTC or thioredoxin a rapid and transient increase of *c-jun* and *c-fos* mRNA levels was detected (Meyer et al., 1993; Schenk et al., 1994). The antioxidant-induced pathway therefore differs from the prooxidant and TPA-induced pathway that may primarily involve posttranslational regulation of AP-1.

The increased expression of *c-fos* by antioxidants is mediated by the SRE (Meyer et al., 1993). The SRE is constitutively occupied by a dimer of the serum response factor (SRF) and an associated transcription factor belonging to the *ets* proto-oncogene family, called ternary complex factor (TCF) or Elk-1 (reviewed in Janknecht and Nordheim, 1993). Recent studies have shown that the transactivation domain of TCF is phosphorylated by MAP kinase(s) in response to serum stimulation (Gille et al., 1992; Zinck et al., 1993; Marais et al., 1993). This modification of TCF is required to activate the transactivating potential of SRF. We have found that both hydrogen peroxide and various antioxidants, like serum and TPA stimulation, induce phosphorylation of TCF by the activation of MAP kinases (J. Müller et al., 1997). Hence, prooxidant as well as antioxidant signals appear to converge at the level of MAP kinases and activate them. The mechanism by which prooxidant and antioxidant stimuli activate MAP kinase remains to be established. Although both signals activate the SRE and lead to *c-fos* expression, highly active AP-1 is only found after antioxidant stimulation. Therefore, additional posttranscriptional control levels seem to be imposed allowing full AP-1 activation only under antioxidant or hypoxic conditions.

At first view, it seemed weird that contrary stimuli such as prooxidants and antioxidants induce similar patterns of gene expression. However, it has been reported that *gadd 153*, a gene of still unknown function, is highly induced upon incubation with toxic agents and H_2O_2, but is also expressed upon treatment with the reductant dithiothreitol (Chen et al., 1992). In the genes coding for glutathione S-transferase (Ya subunit) and NAD(P)H:quinone reductase a DNA sequence, called antioxidant-responsive element (ARE), has been identified that reveals a strong sequence homology with the classical AP-1 consensus sequence (Rushmore et al., 1991; Friling et al., 1992; Li and Jaiswal, 1992). Both genes are responsive to phenolic antioxidants, such as BHA, but can also be induced by hydrogen peroxide. This suggests that cells can respond to a shift to either prooxidant or antioxidant conditions by activating the same kinase, the same transcription factor and, ultimately, the same or similar pattern of gene expression.

VII. DNA DAMAGE AS AN INTERMEDIATE IN OXIDANT-INDUCED GENE EXPRESSION?

Oxidative stress and UV irradiation, both inducers of NF-κB- and AP-1-dependent gene expression, are known to cause DNA damage by forming pyrimidine dimers and other DNA lesions (Imlay and Linn, 1988). Therefore, it has been speculated that, like bacteria, eukaryotes contain a kind of "Rec A response," which is triggered by damaged DNA or modified nucleotides (reviewed in Holbrook and Fornace, 1991; Herrlich et al., 1992). This idea was supported by the observation that a considerable overlap exists between the UV action spectrum on AP-1-dependent gene activation and the UV absorbance by DNA (Stein et al., 1989; Mai et al., 1989). Furthermore, in cells from patients with Xeroderma pigmentosum which are deficient in repair of pyrimidine dimers, a much lower dose of UV is required to increase transcription of AP-1-dependent genes. If DNA lesions are indeed required for UV-induced gene activation, at least in the case of cytoplasmic NF-κB, a reflow of signals from the nucleus to the cytoplasm has to be postulated. However, Devary et al. (1993) demonstrated that enucleated cells induce NF-κB equally well in response to TPA or UV. This observation suggests that cytoplasmic rather than nuclear DNA-dependent events are involved in triggering UV-induced transcriptional activation.

VIII. ROLE OF TYROSINE PROTEIN KINASES AND OTHER DOWNSTREAM SIGNALS

Various membrane-bound and cytoplasmic protein kinases have recently been reported to play a role in transducing signals towards NF-κB and AP-1 (Devary et al., 1992; 1993; Radler-Pohl et al., 1993; Finco and Baldwin, 1993; Li and Sedivy, 1993). Binding of growth factors or cytokines to their membrane receptors is frequently accompanied by autophosphorylation of their receptors or association with tyrosine kinases. Important components which couple extracellular signals to downstream targets are the membrane-attached protein tyrosine kinases of the Src family. UV-irradiation also rapidly activates the tyrosine kinase activity of c-Src, an event followed by the activation of the membrane-bound GTP-binding protein Ha-Ras and the cytoplasmic serine/threonine-specific kinase Raf-1 (Devary et al., 1992; 1993). Using transfection of transdominant-negative mutants, evidence has been obtained that this signal pathway is involved in UV-mediated activation of NF-κB as well as AP-1 (Devary et al., 1992, 1993). In cells transfected with c-src and Ha-ras mutants, activation of AP-1 and NF-κB was significantly suppressed. Furthermore, inhibition of tyrosine kinase activity by herbimycin prevented induction of the transcription factors in response to UV irradiation. However, induction by phorbol esters was not affected implying that UV irradiation and phorbol esters use distinct proximal pathways. A proposed sequence of events involved in AP-1 activation is shown in Figure 2. The activation of Src kinase was suppressed by the antioxidants N-acetylcysteine and reduced glutathione, suggesting that membrane-associated tyrosine kinases (or phosphatases) serve as sensors for the redox status of the cell (Devary et al., 1993).

Additional evidence supports that oxidative stress activates certain tyrosine kinases. The transmembrane kinase LTK which is located in the endoplasmic reticulum becomes autophosphorylated upon exposure to the oxidant diamide (Bauskin et al., 1991). In particular, in T and B lymphocytes, Src family kinases have been implicated in proximal events of cell activation. Recent studies have identified that two Src members, $p56^{lck}$ and $p59^{fyn}$, are activated by hydrogen peroxide and diamide (Schieven et al., 1993; Nakamura et al., 1993). It appears that activation of Src tyrosine kinases is the earliest detectable event following treatment of cells with stimuli causing prooxidant conditions.

Since NF-κB and AP-1 are different with respect to subcellular localization and mode of activation, it is obvious that an initially common pathway must diverge in order to cause distinct downstream events. Activation of Raf-1 kinase may still be a step shared between NF-κB and AP-1 activation in response to treatments with

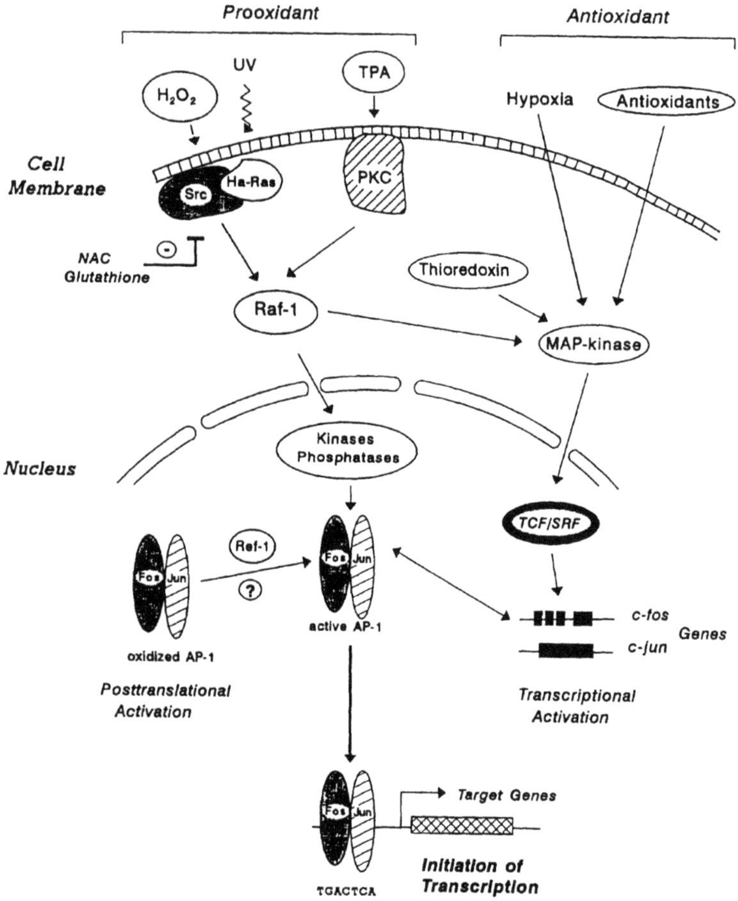

Figure 2. Posttranslational and transcriptional activation of AP-1 under prooxidant and antioxidant conditions: Prooxidants such UV irradiation and hydrogen peroxide activate AP-1 primarily by posttranslational mechanisms. This process involves phosphorylation and dephosphorylation reactions of the preexisting AP-1 subunits c-Fos and c-Jun. AP-1 activation is further maintained by a positive autoregulatory mechanism involving transcriptional activation of the *c-fos* and *c-jun* gene. The posttranslational modifications are preceded by the activation of membrane-associated Src kinase and Ha-Ras GTP-binding protein, followed by the activation of cytoplasmic Raf-1 kinase. Further activating steps may include the reduction of conserved cysteine residues in c-Fos and c-Jun which can be catalyzed by nuclear Ref-1 protein. Antioxidant or hypoxic conditions, in contrast, induce AP-1 activation entirely by transcriptional activation of *c-fos* and *c-jun* gene expression. Intracellularly, this step may be controlled by thioredoxin. Transcriptional activation of *c-fos* involves activation of MAP-kinases, followed by the phosphorylation of ternary complex factor (TCF) at the SRE within the c-fos promotor.

phorbol ester and other oxdative stress-inducing agents (Devary et al., 1993; Radler-Pohl et al., 1993; Finco and Baldwin, 1993). Since activation of AP-1 involves specific phosphorylation and dephosphorylation reactions of c-Jun, certain nuclear kinases and phosphatases have to be implicated. In contrast, NF-κB is activated in the cytoplasm via dissociation of IκB. Although various cell-free experiments have shown phosphorylation and release of IκB by protein kinase C, protein kinase A, haem-regulated kinase and Raf-1 (Shirakawa and Mizel, 1989; Ghosh and Baltimore, 1990; Li and Sedivy, 1993), the *in vivo* role of IκB phosphorylation and the identity of the involved kinase(s) are presently unclear. Instead, it has been shown that in intact cells NF-κB activation is associated with a coincident loss of IκB by proteolytic degradation in response to various inducing agents (Henkel et al., 1993). The stimulated degradation of IκB is a necessary reaction because its inhibition by various chymotrypsin inhibitors prevented activation of NF-κB. Three recent reports further demonstrate that the loss of IκB is preceded by its phosphorylation (Beg et al., 1993; Brown et al., 1993; Cordle et al., 1993). Very recent experiments from our laboratory suggest that phosphorylation of IκB is not sufficient for release and activation of NF-κB. A particular protease inhibitor caused an accumulation of phosphorylated IκB following TNF stimulation without subsequent activation of NF-κB (B. Traenckner and P. A. Baeuerle, unpublished results). Although IκB phosphorylation seems to be insufficient for full inactivation of IκB in intact cells, it might still be a necessary reaction which, for instance, tags IκB for proteolytic degradation within its complex with NF-κB.

As described above, Src kinase activation can be blocked by antioxidants, such as N-acetyl-cysteine. This pathway is involved in NF-κB and AP-1 activation in response to UV irradiation, but not to TNF or phorbol ester treatment (Devary et al., 1993). The antioxidant PDTC also prevented IκB degradation in response to TPA suggesting that this more downstream event may involve a further redox-controlled component. Because induction of NF-κB by all inducers tested so far is inhibited by antioxidants, the major target for the inhibitory action of antioxidants must be step further downstream, perhaps at the level of the protease or another kinase.

IX. REDOX-CONTROLLED REGULATION OF DNA BINDING AS A SECOND LEVEL OF CONTROL?

The findings on NF-κB and AP-1 have stimulated investigations on the effects of oxidation/reduction of other transcription factors. In this way, it has been observed that, for instance, the heat shock transcription factor HSF is also activated under prooxidant conditions (Bruce et al., 1993). In contrast to the activation of factors by prooxidants, several studies have shown that under cell-free conditions the DNA-binding activity of many transcription factors requires antioxidants. The DNA-binding activity of NF-κB and AP-1 is eliminated after treatment of cell extracts with the sulfhydryl-modifying agents diamide or N-ethylmaleimide (Toledano and Leonard,

1991; Molitor et al., 1991; Abate et al., 1990), although it its induced by prooxidants in intact cells. A requirement for reducing conditions has also been reported for the DNA-binding activity of c-Myb (Guehmann et al., 1992; Myrset et al., 1993), TFIII C (Cromlish and Roeder, 1989), ISGF3 (Levy et al., 1989), Ets (Wasylyk and Wasylyk, 1993), p53 (Hainaut and Milner, 1993), Egr-1 (Huang and Adamson, 1993) and the glucocorticoid receptor (Grippo et al., 1985). The leucine-zipper molecules c-Fos, c-Jun, CREB-1, but also c-Myb and NF-κB, which bind DNA by other motifs, all contain a cysteine residue flanked by basic amino acids in their DNA binding domain that is critical for DNA binding (Abate et al, 1990; Guehmann et al., 1992; Kumar et al., 1992; Xanthoudakis et al., 1992). Since these cysteine residues are highly conserved and their reduced state is essential for DNA binding, it has been speculated that redox modification of DNA binding may be another physiological mechanism of controlling transcriptional activity. However, all present data rely on cell-free experiments and no direct evidence for the existence of oxidized cysteine residues in transcription factors within intact cells has been obtained so far. There is also a vast literature showing that enzymatic activities of cytosolic enzymes are lost due to cysteine oxidation when proteins become exposed to oxidizing conditons upon isolation from their highly reducing subcellular compartments (reviewed in Ziegler, 1985).

There are a few indications for a physiological role of direct redox control. For instance, it has been noted that the *v-jun* oncogene has the critical cysteine in the DNA binding domain replaced by a serine (Chida and Vogt, 1992). This substitution, which enhances the transforming potential of the oncogene product, would escape redox modification. Recently, a ubiquitous nuclear protein, called Ref-1, has been shown to enhance the DNA-binding activity of Fos, Jun and some other factors in cell-free systems (Xanthoudakis et al., 1992). Apart from being a DNA repair enzyme, Ref-1 catalyzes through a separate protein domain the reduction of thiol groups by the transfer of electrons from reduced thioredoxin. At present, no evidence is available that Ref-1 is involved in the intranuclear redox control of transcription factors. A physiological role would imply that either Ref-1 activity or the redox state inside the nucleus is modified under certain circumstances but no evidence exists for either possibility. Instead, we have observed that even upon treatment of cells with high doses of prooxidants the redox state within the nucleus is kept fairly constant since no significant changes in the ratio of reduced to oxidized glutathione are observed (M. Los and K. Schulze-Osthoff, unpublished results). In contrast, several NF-κB-inducing agents lead to a rapid and transient decrease of reduced glutathione in the cytoplasm. Therefore, it can be expected that redox regulation of transcription factors occurs in the cytoplasmic rather than in the nuclear compartment.

X. ROLE OF ROIs IN TNF-MEDIATED SIGNALING

Recent studies have suggested that gene induction by proinflammatory cytokines, such as IL-1 and TNF, may involve intracellular formation of ROIs. TNF was originally described based on its cytotoxic activity for tumor cells (reviewed in Fiers, 1991). Later on, the cytokine has been identified as an important activator of several immunoregulatory processes due to its capacity to induce a great variety of immunorelevant genes. It now emerges that TNF is a cardinal mediator of several diseases in which ROIs play an established role. These include allograft rejection, septic shock, chronic inflammatory diseases, ischemia/reperfusion syndrome, acute respiratory distress syndrome, silicosis and various other respiratory disorders (reviewed in Vasalli, 1992). TNF causes oxidative stress by various enzymatic mechanisms which may contribute to its potent NF-κB-inducing effect. In neutrophils, TNF activates the membrane-bound NADPH oxidase that can generate large amounts of ROI in the oxidative burst (Rossi, 1986). In endothelial cells, oxidative stress induced by TNF has largely been attributed to the activation of cytosolic xanthine oxidase which generates O_2^- and H_2O_2 during the metabolism of xanthine or hypoxanthine to uric acid (Friedl et al., 1989). This enzyme normally exists as a xanthine dehydrogenase but can be proteolytically cleaved into an ROI-producing oxidase form. Recent studies suggest that ROI formation in the respiratory chain of mitochondria is an important event for the cytotoxicity as well as for NF-κB mediated gene induction by TNF.

In L929 fibrosarcoma cells, both cytotoxicity and NF-κB induction were blocked by inhibitors preventing ROI formation in the mitochondrial respiratory chain (Schulze-Osthoff et al., 1992; 1993). Studies using isolated mitochondria have established that reduced ubiquinone in the respiratory chain may be a major source of ROI production (Boveris et al., 1976; Turrens et al., 1985). NF-κB induction by TNF is strongly inhibited by rotenone and amytal, two drugs that inhibit the electron transfer at the level of complex I, thereby preventing ROI formation at the distally located ubiquinone (Schulze-Osthoff et al., 1992, 1993). In contrast, antimycin A, a drug that acts distally from ubiquinone, potentiates TNF-induced NF-κB activation. This might occur by an increase of the reduced form of ubiquinone and consequent increase of superoxide anion formation at this site (Konstantinov et al., 1987; Cino and DelMaestro, 1989). These observations suggest that a signal delivered through the 55kD-TNF receptor has to communicate with the mitochondrial proteins to induce ROI formation by leakage of electrons from the electron transfer chain. Further evidence for a crucial role of mitochondria-derived ROIs is the finding that in subclones of L929, which are devoid of mitochondrial DNA, TNF cytotoxicity and NF-κB activation are substantially reduced (Schulze-Osthoff et al., 1993). Packer and coworker demonstrated that particularly water-soluble esters of vitamin E are much more potent inhibitors of NF-κB activation than lipophilic vitamin E (Suzuki and Packer, 1993). This suggests that these antioxidants are not exerting their inhibitory effect at or within the plasma membrane, but instead have to pass through into an aqueous compartment such as the cytosol. Radical sources other than mitochondria seem to play only a minor role for NF-κB activation by TNF in L929 cells. Though these cells contain NADPH oxidase activity similar to untransformed fibroblasts, inhibition of this enzyme did not affect

TNF-induced NF-κB activation (K. Schulze-Osthoff and B. Meier, unpublished results; Meier et al., 1989). Likewise allopurinol, an inhibitor of cytoplasmic xanthine oxidase did not block activation of NF-κB (Schulze-Osthoff et al., 1993). However, since NF-κB is induced by various distinct stimuli, NF-κB activation may involve ROI formation by more than one mechanism.

The available data on TNF signaling provide an illustrating example that also eukaryotes contain an adaptive program confering protection against various forms of oxidative stress. In various cell types, TNF selectively induces the gene encoding manganeous superoxide dismutase (MnSOD) (Wong and Goeddel, 1988; Visner et al., 1990). Overexpression of this enzyme provides increased resistance towards TNF cytotoxicity (Wong et al., 1989). Consequently, cells which are pretreated with sublethal doses of TNF become resistant towards higher doses that would be otherwise lethal. Resistance can also be induced by transfection of the TNF gene into sensitive cell lines. Furthermore, rats that are pretreated with TNF before exposure to hyperoxia, show less pulmonary injury (White et al., 1987). It has been found that this adaptive response to oxidative stress is associated with the expression of several antioxidant enzymes in lungs. In addition to MnSOD, TNF also upregulates the expression of thioredoxin, ferritin and metallothionein, all proteins with protective functions (White et al., 1989; Liu et al., 1991; Yodoi and Uchiyama, 1992).

XI. ACTIVATION OF OXIDATIVE STRESS-INDUCIBLE GENES

In this article we have described the redox regulation of the transcription factors NF-κB and TNF by various forms of oxidative stress. Both factors have a great number of target genes which are all likely to be induced upon changes in the intracellular redox state. Space does not permit us to cover all genes, but the reader is referred to detailed reviews (Baeuerle, 1991; Baeuerle and Henkel, 1994; Grilli et al., 1993, Herrlich et al., 1992; Holbrook and Fornace, 1991) and the following chapters of this volume. In contrast to bacteria, the mammalian response to ROIs involves also the induction of genes associated with cellular processes other than defense (Table 1). These processes include signal transduction (growth factors, growth factor receptors, transcription factors), proliferative events (oncogenes), tissue injury (proteases) and inflammatory processes (cytokines, cytokine receptors, cell adhesion molecules). NF-κB has a central role in gene control associated with immunological defense mechanisms. For instance, the expression of endothelial cell adhesion molecules is coordinately regulated by NF-κB in response to cytokines. It is thus not unexpected to find that the genes coding for ELAM-1, VCAM-1 and ICAM-1 are all induced by hydrogen peroxide (Lo et al., 1993; Marui et al., 1993; Suzuki et al., 1992) as well. This finding might be of relevance for the massive tissue infiltration of leukocytes in inflammatory and oxidative stress-

Table 1. Gene Expression Induced by Oxidative Stress or Inhibited by Antioxidants

Class	*Target gene*	*Reference*
Viruses		
	Human immunodeficiency virus 1 (HIV-1)	Roederer et al., 1990, Mihm et al., 1991; Schreck et al., 1991
	Simian virus 40 (SV40)	Garaci et al., 1992
Cell adhesion molecules		
	Endothelial leukocyte adhesion molecule-1 (ELAM-1)	Suzuki et al., 1992 Schulze-Osthoff et al., 1994
	Vascular cell adhesion molecule-1 (VCAM-1)	Marui et al., 1993 Schulze-Osthoff et al., 1994
	Intercellular cell adhesion molecule-1 (ICAM-1)	
Cytokines and growth factors		
	Interleukin-1 (IL-1)	Koga et al., 1992 Shreeniwas et al., 1992
	Interleukin-2 (IL-2)	Roth & Dröge, 1987; 1991 Los et al., 1995
	Interleukin-6 (IL-6)	Brach et al., 1993 Schulze-Osthoff et al., 1993; 1994
	Interleukin-8 (IL-8)	DeForge et al., 1992; 1993
	Colony stimulating factor-1 (CSF-1)	Satriano et al., 1993
	Tumor necrosis factor α (TNFα)	Hallahan et al., 1989 Sherman et al., 1991
	Monocyte chemotactic protein 1 (JE/MCP-1)	Nose et al., 1991 Satriano et al., 1993
	Platelet-derived growth factor (PDGF)	Witte et al., 1989
Transcription factors		
	c-myc	Crawford et al, 1988; Amstad et al., 1992
	c-jun	Sherman et al., 1990; Nose et al., 1991 Devary et al., 1991
	jun B	Devary et al., 1991
	c-fos	Crawford et al., 1988; Shibanuma et al., 1990, Nose et al., 1991
	egr-1	Nose et al., 1991; Datta et al., 1992
Others		
	Heme oxygenase	Keyse & Tyrrell, 1989; Applegate et al., 1991

continued

Table 1. (Continued)

Class	*Target gene*	*Reference*
Others		
	Glutathione-S-transferase Ya	Rushmore et al. 1991; Friling et al., 1992
	NAD(P)H:quinone oxidoreductase	Li & Jaiswal, 1992
	β-DNA polymerase	Fornace et al., 1989a; Kedar et al., 1991
	Ornithin decarboxylase	Marsh & Mossman, 1991
	Protein tyrosine phosphatase	Keyse & Emslie, 1992
	Metallothionein	Stein et al., 1989
	Collagenase	Stein et al., 1989
	gadd 45	Fornace et al., 1988, 1989b
	gadd 153	Fornace et al., 1988, 1989b

Note: For a detailed compilation of genes induced by UV and DNA-damaging agents the reader is referred to Holbrook & Fornace, 1991; Herrlich et al., 1992.

associated processes. In endothelial and other cell types, various cytokines and growth factors are induced by oxidative stress. Examples are IL-1 (Koga et al., 1992, Shreeniwas et al., 1992), IL-2 (Roth & Dröge, 1987), IL-6 (Brach et al., 1993; Schulze-Osthoff et al., 1994), IL-8 (DeForge et al., 1992; 1993), TNF (Hallahan et al., 1989; Sherman et al., 1991), PDGF and FGF (Witte et al., 1989), MCP-1 and colony-stimulating factors (Satriano et al., 1993). ROI-induced gene expression may be of high relevance for all diseases associated with enhanced formation of ROIs, including septic shock, ischemia/reperfusion and chronic inflammatory diseases (Halliwell and Gutteridge. 1990). ROI-inducible cytokines, such as TNF and IL-1, are themselves known to induce ROI formation, suggesting a positive autoregulatory circuit under these pathological conditions.

Great attention is being paid to the redox control of human immunodeficiency virus (HIV) replication (reviewed in Dröge et al., 1992). Proviral expression and virus replication is largely controlled by binding of NF-κB to two highly conserved regulatory elements in the HIV enhancer/promoter (Nabel and Baltimore, 1987; Duh et al., 1989; Osborn et al., 1989). It has been found that proviral expression is inhibited by several antioxidants such as N-acetyl-L-cysteine, L-cysteine and vitamin C (Roederer et al., 1990; Staal et al., 1990; Mihm et al., 1991). There is a further link of HIV expression to prooxidant conditions. Asymptomatic HIV-infected individuals reveal significantly reduced levels of intracellular glutathione and plasma thiols suggesting that a disturbance of the intracellular redox balance might be important in the pathogenesis of AIDS (Dröge et al., 1988; Eck et al., 1989; Buhl et al., 1989; Staal et al., 1992). Several authors have therefore proposed to administer antioxidant compounds, such as N-acetylcysteine, vitamin C, vitamin E derivatives or other antioxidant substances as an adjuvant therapy in AIDS patients (Dröge et al., 1992; DeQuay et al., 1992; Roederer et al., 1993).

There is good evidence to suggest that ROIs and ROI-controlled transcription factors are also implicated in the regulation of proliferative events. Although high amounts of ROIs are certainly cytotoxic and lead to cell death, sublethal doses of hydrogen peroxide can be mitogenic for certain cell types. Micromolar amounts of hydrogen peroxide stimulate T cell proliferation (Roth & Dröge, 1987). In adipocytes, H_2O_2 exerts insulinomimetic actions resulting in enhanced cell proliferation (Heffetz and Zick, 1989). Treatment with H_2O_2 activates insulin receptor tyrosine kinase activity by inducing autophosphorylation of its β subunit (Heffetz and Zick, 1989). As has been described above, prooxidant conditions may also activate Src tyrosine kinases and thereby modulate intracellular signaling associated with proliferation and immediate-early gene expression. Various members of the NF-κB family seem to be involved in the control of cell proliferation. v-rel, p65Δ, Bcl-3 and lyt-10 were reported to have oncogenic potential (Gilmore, 1991; Narayanan et al., 1992; Ohno et al., 1990; Neri et al., 1991). A correlation between prooxidant stimulation, NF-κB activation and cell proliferation is found in fibroblasts. Low doses of ROI stimulate fibroblast proliferation (Murrell et al., 1990; Shibanuma et al., 1990). In 3T3 fibroblasts, NF-κB activation has been observed to occur specifically during the G_0/G_1 transition, a time point when various immediate-early genes are induced (Baldwin et al., 1991). The *c-myc* proto-oncogene is apparently activated by NF-κB upon serum stimulation (Duyao et al., 1990). We have recently observed that 50 μM H_2O_2 strongly increase proliferation of ESb-L T lymphoma cells (Los et al., 1994). This is also accompanied by an increased NF-κB activation and expression of the *IL-2* gene. Certainly, reducing conditions may be beneficial for long term growth in most cell types. Nonetheless, ROIs may be important to initiate certain early events associated with growth control. It is possible that a primary pro-oxidative stimulus intiates the synthesis of glutathione, thioredoxin, or other protective molecules which then shift the intracellular redox state to reducing conditions being more beneficial for DNA synthesis. It is well established that AP-1 is crucially involved in the control of cell cycle and proliferation. The expression of the proto-oncogenes *c-fos* and *c-jun* as well as *c-myc* is greatly enhanced by H_2O_2 (Crawford et al., 1988; Amstad et al., 1992) but activation of the resulting factor AP-1 might only be optimal under the subsequently induced reducing conditions (Meyer et al., 1993; Schenk et al., 1994). Antioxidant or hypoxic conditions as they exist in tumors may therefore support proliferation by allowing optimal AP-1 activation. Since AP-1 controls the expression of several proteases such as collagenase, stromelysin and plasminogen activator, antioxidant conditions are also likely to be involved in tumor promotion by supporting the spreading of tumor cells in tissues (Cerutti, 1985, 1991; Trush and Kensler, 1991).

In contrast to low amounts of ROIs, which may stimulate proliferation by producing an ROI-protective environment, high amounts of ROIs lead to growth arrest and cell death. A delay of the cell cycle at the G_2 checkpoint is a common response to oxidative stress and requires the synthesis of new gene products to overcome it (Janssen et al., 1993). It is thought that the delay in G_2 has a protective effect, presumably, by allowing DNA damage to be repaired prior to mitosis. Several transcripts have been identified which are expressed at this checkpoint and are induced by hydrogen peroxide, DNA-damaging agents and other growth cessation signals (reviewed in Holbrook and Fornace, 1991). The genes encoding these transcripts have been designated gadd (growth arrest and DNA damage-inducible, Fornace et al., 1988, 1989b). Members of this family represent a new class of molecules that may be involved in negative control of cell growth.

Unlike many other stress-inducible genes, gadd 45 its not induced by phorbol esters and its promoter does not contain an AP-1- or NF-κB-like site. The gene coding for heme oxygenase, an enzyme thought to play a protective role during oxidative stress, is also highly induced by oxidative stress (Keyse and Tyrrell, 1989; Applegate et al., 1991). Since the *cis*-regulatory elements responsible for the oxidant-triggered induction of these genes have not yet been defined, it is likely that novel transcription factors will be identified which are selectively induced by oxidative stress.

XII. CONCLUSIONS

ROIs are frequently regarded as toxic and therefore harmful metabolites implicated in the deleterious effects of several diseases. Accumulating evidence now exists that low concentrations of ROIs may provide a versatile cellular control mechanism and play an important role in gene regulation. ROI-mediated gene induction is apparently not restricted to situations causing acute oxidative stress, such as exposure to oxidizing agents, UV light and ionizing irradiation. As has been illustrated in the case of NF-κB, ROIs may be produced in response to physiological inducers and act as second messenger molecules in intracellular signal transduction. ROIs appear as suitable canditates for this role as they are extremely short-lived molecules, which allows a tight control of their formation and degradation by a machinery of different enzymes and low molecular weight antioxidants. Because many physiological stimuli, such as TNF, seem to utilize ROIs for signal transduction, it is not suprising to find a considerable overlap in the pattern of genes induced by ROIs and cytokines. In apparent contrast to bacteria, ROI-inducible gene expression in higher eukaryotes is not exclusively used to adapt cells for survival in a prooxidant environment. Only a few of the eukaryotic genes induced by ROI-dependent signals have ROI-protective functions. Most genes are rather involved in signaling processes which are important during inflammation, injury and infection. Hence, it seems that during evolution cells have learnt not only to simply counteract ROIs, but, on one side, to induce oxidative stress in a controlled fashion and, on the other side, to use ROIs as signaling molecules in a great variety of normal regulatory processes.

ACKNOWLEDGMENTS

We are grateful to Dr. Heike Pahl and Marek Los for their critical reading of this manuscript.

REFERENCES

Abate, C., Patel, L., Rauscher, F. J. III, & Curran, T. (1990). Redox regulation of Fos and Jun DNA-binding activity *in vitro*. Science 249, 1157-1161.

Amabile-Cuevas, C. F., & Demple, B. (1991). Molecular characterization of the *soxRS* genes of *Escherichia coli*: Two genes control a superoxide stress regulon. Nucleic Acids Res. 19, 4479-4484.

Amstad, P. A., Peskin, A., Shah, G., Mirault, M. E., Moret, R., Zbinden, I., & Cerutti, P. A. (1991). The balance between Cu/Zn-superoxide dismutase and catalase affects the sensitivity of mouse epidermal cells to oxidative stress. Biochemistry 30, 9305-9313.

Amstad, P.A., Krupitza, G., & Cerutti, P. A. (1992). Mechanism of *c-fos* induction by active oxygen. Cancer Res. 52, 3952-3960.

Angel, P., Hattori, K., Smeal, T., & Karin, M. (1988). The jun proto-oncogene is positively autoregulated by its product, Jun/AP-1. Cell 55, 875-885.

Angel, P., & Karin, M., (1991). The role of Jun, Fos, and the AP-1 complex in cell proliferation and transformation. Biochem. Biophys. Acta 1072, 129-157.

Applegate, L. A., Luscher, P., & Tyrrell, R. M. (1991). Induction of heme oxygenase: A general response to oxidant stress in cultured mammalian cells. Cancer Res. 51, 974-978.

Baeuerle, P. A. (1991). The inducible transcription activator NF-κB: regulation by distinct protein subunits. Biochim. Biophys. Acta 1072, 63-80.

Baeuerle, P. A., & Henkel, T. (1994). Function and activation of NF-κB in the immune system. Ann. Rev. Immunol. 12, 141-179.

Baeuerle, P. A., & Baltimore, D. (1991). The physiology of the NF-κB transcription factor. In: Molecular Aspects of Cellular Regulation, Hormonal Control Regulation of Gene Transcription. (Cohen, P. and Foulkes, J. G., eds.), pp. 409-432. Elsevier/North Holland Biomedical.

Baeuerle, P. A., & Baltimore, D. (1988). A specific inhibitor of the NF-κB transcription factor. Science 242, 540-546.

Baldwin, A. S., Azizkhan, J. C., Jensen, D. E., Beg, A. A., & Coodly, L. R. (1991). Induction of NF-κB DNA-binding activity during the G_0 to G_1 transition in mouse fibroblasts. Mol. Cell Biol. 11, 4943-4951.

Bauskin, A. R., Alkalay, I., & Ben-Neriah, Y. (1991). Redox regulation of a protein tyrosine kinase in the endoplasmic reticulum. Cell 66, 685-696.

Beg, A. A., & Baldwin, A. S. Jr. (1993). The IκB proteins: multifunctional regulators of Rel/NF-κB transcription factors. Genes & Dev. 7, 2064-2070.

Beg, A. A., Finco, T. S., Nantermet, P. V., & Baldwin, A. S. Jr. (1993). Tumor necrosis factor and interleukin-1 lead to phosphorylation and loss of IκBα: a mechanism for NF-κB activation. Mol. Cell. Biol. 13, 3301-3310.

Binétruy, B., Smeal, T., Meek, J., & Karin, M. (1991). Ha-Ras augments c-Jun activity and stimulates phosphorylation of its activation domain. Nature 351, 122-127.

Blank, V., Kourilsky, P., & Israël, A. (1992). NF-κB and related proteins: Rel/dorsal homologies meet ankyrin-like repeats. Trends Biochem. Sci. 17, 135-140.

Bomtsztyk, K., Rooney, J. W., Iwasaki, T., Rachie, N. A., Dower, S. K., & Sibley, C. H. (1991). Evidence that interleukin-1 and phorbol esters activate NF-κB by different pathways: role of protein kinase C. Cell Regulation 2, 329-335.

Boveris, A., Cadenas, E., & Stroppani, A. O. M. (1976). Role of ubiquinone in the mitochondrial generation of hydrogen peroxide. Biochem. J. 156, 435-445.

Boyle, W. J., Smeal, T., Defize, L. H. K., Angel, P., Woodgett, J. R., Karin, M., & Hunter, T. (1991). Activation of protein kinase C decreases phosphorylation of *c-jun* at sites that negatively regulate its DNA-binding activity. Cell 64, 573-584.

Brach, M. A., Gruβ, H.-J., Kaisho, T., Asano, Y., Hirano, T., & Herrmann, F. (1993). Ionizing radiation induces expression of interleukin 6 by human fibroblasts involving activation of nuclear factor-κB. J. Biol. Chem. 268, 8466-8472.

Brown, K., Park, S., Kanno, T., Franzoso, G., & Siebenlist, U. (1993). Mutual regulation of the transcriptional activator NF-κB and its inhibitor, IκB-α. Proc. Natl. Acad. Sci. USA 90, 2532-2536.

Bruce, J. L., Price, B. D., Coleman, N., & Calderwood, S. K. (1993). Oxidative injury rapidly activates the heat shock transcription factor but fails to increase levels of heat shock proteins. Cancer Res. 53, 12-15.

Buhl, R., Jaffe, H. A., Holroyd, K. J., Wells, F. B., Mastangeli, A., Saltini, C., Cantin, A. M., & Crystal, R. G. (1989). Systemic glutathione deficiency in symptom-free HIV-seropositive individuals. Lancet 1294-1296.

Büscher, M., Rahmsdorf, H. J., Litfin, M., Karin, M., & Herrlich, P. (1988). Activation of c-fos gene by UV and phorbol ester: different signal transduction pathways converge to the same enhancer element. Oncogene 3, 301-311.

Cerutti, P. A. (1985). Prooxidant states and tumor promotion. Science 227, 375-381.

Cerutti, P. A. (1991). Oxidant stress and carcinogenesis. Eur. J. Clin. Invest. 21, 1-5.

Chen, Q., Yu, K., Holbrook, N. J., & Stevens, J. L. (1992). Activation of the growth arrest and DNA damage-inducible gene gadd 153 by nephrotoxic cysteine conjugates and dithiothreitol. J. Biol. Chem. 267, 8207-8212.

Chida, K., & Vogt, P. K. (1992). Nuclear translocation of viral Jun but not of cellular Jun is cell cycle dependent. Proc. Natl. Acad. Sci. USA 89, 4290-4294.

Cino, M., & Del Maestro, R. F. (1989). Generation of hydrogen peroxide by brain mitochondria: the effect of reoxygenation following postdecapitative ischemia. Arch. Biochem. Biophys. 269, 623-638.

Cordle, S. R., Donald, R., Read, M. A., & Hawinger, J. 1993. Lipopolysaccharide induces phosphorylation of MAD3 and activation of c-Rel and related NF-κB proteins in human moncoytic THP-1 cells. J. Biol. Chem. 268, 11803-11810.

Crawford, D., Zbinden, J., Amstad, P., & Cerutti, P. A. (1988). Oxidant stress induces the proto-oncogenes *c-fos* and *c-myc* in mouse epidermal cells. Oncogene 3, 27-32.

Cromlish, J. A., & Roeder, G. (1989). Human transcription factor IIIC (TFIIIC). J. Biol. Chem. 264, 18100-18109.

Datta, R., Hallahan, D. E., Kharbanda, S. M., Rubin, E., Sherman, M. L., Huberman, E., Weichselbaum, R. R., & Kufe, D. W. (1992). Involvement of reactive oxygen intermediates in the induction of the *c-jun* transcription by ionizing radiation. Biochemistry 31, 8300-8306.

DeQuay, B., Malinverni, R., & Lauterburg, B. H. (1992). Glutathione depletion in HIV-infected patients: role of cysteine deficiency and effect of oral N-acetylcysteine. AIDS. 6, 815-819.

DeForge, L. E., Preston, A. M., Takeuchi, E., Kenney, J,. Boxer, L., & Remick, D. R. (1993). Regulation of interleukin 8 expression by oxidant stress. J. Biol. Chem. 268, 25568-25576.8

DeForge, L. E., Fantone, J. C., Kenney, J. S., & Remick, D. (1992). Oxygen radical scavengers selectively inhibit interleukin 8 production in human whole blood. J. Clin. Invest. 90, 2123-2129.

Demple, B., & Amabile-Cuevas, C. F. (1991). Redox redux: the control of oxidative stress responses. Cell 67, 837-839.

Demple, B. (1991). Regulation of bacterial oxidative stress genes. Annu. Rev. Genet. 25, 315-337.

Devary, Y., Gottlieb, R. A., Laus, L. F., & Karin, M. (1991). Rapid and preferential activation of the c-jun gene during the mammalian UV response. Mol. Cell. Biol. 11, 2804-2811.

Devary, Y., Gottlieb, R. A., Smeal, T., & Karin, M. (1992). The mammalian ultraviolet response is triggered by activation of Src tyrosine kinases. Cell 71, 1081-1091.

Devary, Y., Rosette, C., DiDonato, J. A., & Karin, M. (1993). NF-κB activation by ultraviolet light not dependent on a nuclear signal. Science 261, 1442-1445.

Dominguez, I., Sanz, L., Arenzana-Seisdedos, F., Diaz-Meco, M. T., Virelizier, J. L., & Moscat, J. 1993. Inhibition of protein kinase C ζ subspecies blocks the activation of an NF-κB-like activity in Xenopus oocytes. Mol. Cell Biol. 13, 1290-1295.

Dröge, W., Eck, H.-P., & Mihm, S. (1992). HIV-induced cysteine deficiency and T-cell dysfunction—a rationale for treatment with N-acetylcysteine. Immunol. Today 13, 211-213.

Dröge, W., Eck, H.-P., Näher, H., Pekar, U., & Daniel, V. (1988). Abnormal amino acid concentrations in the blood of patients with acquired immunodeficiency syndrome (AIDS) may contribute to the immunological defect. Biol. Chem. Hoppe-Seyler 369, 143-148.

Duh, E. J., Maury, W. J., Folks, T. M., Fauci, A. S., & Rabson, A. B. (1989). Tumor necrosis factor α activates human immunodeficiency virus type 1 through induction of nuclear factor binding to the NF-κB sites in the long terminal repeat. Proc. Natl. Acad. Sci. USA 86, 5974-5978.

Duyao, M. P., Buckler, A. J. & Sonnenschein, G. E. (1990). Interaction of an NF-κB-like factor with a site upstream of the c-myc promoter. Proc. Natl. Acad. Sci. USA. 87, 4727-4731.

Eck, H.-P., Gmünder, H., Hartmann, H., Petzoldt, D., Daniel, V., & Dröge, W. (1989). Low concentrations of acid soluble thiol (cysteine) in the blood plasma of HIV-1 infected patients. Biol. Chem. Hoppe-Seyler 370, 101-108.

Fiers, W. (1991). Tumor necrosis factor. Characterization at the molecular, cellular and *in vivo* level. FEBS Lett. 285, 199-212.

Finco, T. S., & Baldwin, A. S. (1993). κB site-dependent induction of gene expression by diverse inducers of nuclear factor κB requires Raf-1. J. Biol. Chem. 268, 17676-17679.

Fornace, A. J. J., Zmudska, B. Hollander, M. C., & Wilson, S.H. (1989a). Induction of beta polymerase mRNA by DNA damaging agents in Chinese hamster ovary cells. Mol. Cell. Biol. 9, 851-853.

Fornace, A. J. J., Nebert, D. W., Hollander, M. C., Luethy, J. D., Papathanasiou, M., Fargnoli, J., & Holbrook, N. J. (1989b). Mammalian genes coordinately regulated by growth arrest signals and DNA-damaging agents. Mol. Cell. Biol. 9, 4196-4203.

Fornace, Jr. A. J., Alamo, Jr. I., & Hollander, M. C. (1988). DNA damage-inducible transcripts in mammalian cells. Proc. Natl. Acad. Sci. USA. 85, 8800-8804.

Friedl, H. P., Till, G. O., Ryan, U. S., & Ward, P. A. (1989). Mediator-induced activation of xanthine oxidase in endothelial cells. FASEB J. 3, 2512-2518.

Friling, R. S., Bergelson, S., & Daniel, V. (1992). Two adjacent AP-1-like binding sites form the electrophile-response element of the murine glutathione S-transferase Ya subunit gene. Proc. Natl. Acad. Sci. USA 89, 668-672.

Garaci, E., Palamar, A. T., Di-Francesco, P., Favalli, C., Ciriolo, M. R., & Rotilio, G. (1992). Glutathione inhibits replication and expression of viral proteins in cultured cells infected with Sendai virus. Biochem. Biophys. Res. Commun. 188, 1090-1096.

Ghosh, S., & Baltimore, D. (1990). Activation in vitro of NF-êB by phosphorylation of its inhibitor IκB. Nature 344, 678-682.

Gille, H. J., Sharrocks, A. D., & Shaw, P. E. (1992). Phosphorylation of transcription factor $p62^{TCF}$ by MAP kinase stimulates ternary complex formation at c-fos promotor. Nature, 358, 414-417.

Gilmore, T. D. (1991). Malignant transformation by mutant Rel proteins. Trends Genet. 7, 318-322.

Goebeler, M., Roth, J., Bröcker, E. B., Sorg, C., & Schulze-Osthoff, K., (1995). Activation of nuclear factor-κ B and gene expression in human endothelial cells by the common haptens nickel and cobalt. J. Immunol. 155, 2459-2467.

Grilli, M., Chiu, J. J.-S., & Lenardo, M. J. (1993). NF-êB and Rel: participants in a multiform transcriptional regulatory system. Intern. Rev. Cytol. 143, 1-62.

Grippo, J. F., Holmgren, A., & Pratt, W. B. (1985). Proof that the endogenous heat-stable glucocorticoid receptor activating factor is thioredoxin. J. Biol. Chem. 260, 93-97.

Guehmann, S., Vorbrueggen, G., KalêBrenner, F., & Moelling, K. (1992). Reduction of a conserved Cys is essential for Myb DNA binding. Nucleic Acids Res. 20, 2279-2286.

Hainaut, P., & Milner, J., (1993). Redox modulation of p53 conformation and sequence-specific DNA binding in vitro. Cancer. Res. 53, 4469-4473.

Hallahan, D. E., Spriggs, D. R., Beckett, M. A., Kufe, D. W., & Weichselbaum, R. R. (1989). Increased tumor necrosis factor α mRNA after cellular exposure to ionizing radiation. Proc. Natl. Acad. Sci. USA 86, 10104-10107.

Halliwell, B., & Gutteridge, J. M. C. (1990). The role of free radicals and catalytic metal ions in human disease: an overview. Methods Enzymol. 186, 1-85.

Heffetz, D., & Zick, Y. (1989). H_2O_2 potentiates phosphorylation of novel putative substrates for the insulin receptor kinase in intact Fao cells. J. Biol. Chem. 264, 10126-10132.

Henkel, T., Machleidt, T., Alkalay, I., Krönke, M., Ben-Neriah, Y., & Baeuerle, P. A. (1993). Rapid proteolysis of IκB-α is necessary for activation of transcription factor NF-κB. Nature 365, 182-185.

Herrlich, P., Ponta, H., & Rahmsdorf, H. J. (1992). DNA damage-induced gene expression: signal transduction and relation to growth factor signaling. Rev. Physiol. Biochem. Pharmacol. 119, 187-223.

Hidalgo, E., & Demple, B. (1994). An iron-sulfur center essential for transcriptional activation by the redox-sensing SoxR protein. EMBO J. 13, 138-146.

Holbrook, N. J., & Fornace, A. J., Jr. (1991). Response to adversity: molecular control of gene activation following genotoxic stress. New Biologist 3, 825-833.

Huang, R.-P., & Adamson, E. D. (1993). Characterization of the DNA binding properties of the early growth response-1 (Egr-1) transcription factor: evidence for modulation by a redox mechanism. DNA Cell Biol. 12, 265-273.

Hunter, T., & Karin, M. (1992). The regulation of transcription by phosphorylation. Cell 70, 375-387.

Ignaro, L. I., (1990). Biosynthesis and metabolism of endothelium-derived nitric oxide. Ann. Rev. Parmacol. Toxicol. 30, 535-560.

Imlay, J. A., & Linn, S. (1988). DNA damage and oxygen radical toxicity. Science 240, 1302-1309.

Inoue, J. I., Kerr, L. K., Rashid, D., Davis, N., Bose, H. R., & Verma, I. M. (1992). Direct association of pp40/IκB β with rel/NF-κB transcription factors - role of ankyrin repeats in the inhibition of DNA-binding activity. Proc. Natl. Acad. Sci. USA 89, 4333-4337.

Israël, N., Gougerot-Pocidalo, M.-A., Aillet, F., & Virelizier, J. (1992). Redox status of cells influences constitutive or induced NF-κB translocation and HIV long terminal repeat activity in human T and monocytic cell lines. J. Immunol. 149, 3386-3393.

Janknecht, R., & Nordheim, A. (1993). Gene regulation by Ets proteins. Biochim. Biophys. Acta 1155, 346-356.

Janssen, Y. M. W., Van Houten, B., Borm, P. J. A., & Mossman, B. T. (1993). Biology of Disease. Cell and tissue responses to oxidative damage. Lab. Invest. 69, 261-274.

Kang, S.-M., Tran, A.-C., Grilli, M., & Lenardo, M. (1992). NF-κB subunit regulation in nontransformed CD4+ T lymphocytes. Science 256, 1452-1456.

Karin, M., & Smeal, T. (1992). Control of transcription factors by signal transduction pathways: the beginning of the end. Trends Biochem. Sci. 17, 418-422.

Kedar, P. S., Widen, S. G., Englander, E. W., Fornace, Jr. A. J., & Wilson, S. H. (1991). The ATF-CREB transcription factor-binding site in the polymerase β promotor mediates the positive effect of N-methyl-N'-nitro-N-nitrosoguanidine on transcription. Proc. Natl. Acad. Sci. USA 88, 3729-3733.

Keyse, S. M., & Emslie, E. A. (1992). Oxidative stress and heat shock induce a human gene encoding a protein-tyrosine phosphatase. Nature 359, 644-647.

Keyse, S. M., & Tyrrell, R. M. (1989). Heme oxygenase is the major 32-kDa stress protein induced in human skin fibroblasts by UVA irradiation, hydrogen peroxide, and sodium arsenite. Proc. Natl. Acad. Sci. USA 86, 99-103.

Koga, S., Ogawa, S., Kuwabara, K., Brett, J., Leavy, J. A., Ryan, J., Koga, Y., Plocinski, J., Benjamin, W., Burns, D. K., & Stern, D. (1992). Synthesis and release of interleukin 1 by reoxygenated human mononuclear phagocytes. J. Clin. Invest. 90, 1007-1015.

Konstantinov, A. A., Peskin, A. V., Popova, E. Y., Khomutov, G. B., & Ruuge, E. K. (1987). Superoxide generation by the respiratory chain of tumor mitochondria. Biochim. Biophys. Acta, 894, 1-10.

Kumar, S., Rabson, A. B., & Gélinas, C. (1992). The RxxRxRxxC motif conserved in all Rel/κB proteins is essential for the DNA-binding activity and redox regulation of the v-Rel oncoprotein. Mol. Cell. Biol. 12, 3094-3106.

Lander, H. M., Sehaipal, P., Levine, D. M., & Novogrodsky, A. (1993). Activation of human peripheral blood mononuclear cells by nitric oxide generating compounds. J. Immunol. 150, 1509-1516.

Lee, C. Q., Yun, Y., Hoeffler, J. P., & Habener, J. F. (1990). Cyclic-AMP-responsive transcriptional activation of CREB-327 involves interdependent phosphorylated subdomains. EMBO J. 9, 4455-4465.

Levy, D. E., Kessler, D. S., Pine, R., & Darnell, J. E. Jr. (1989). Cytoplasmic activation of ISGF3, the positive regulator of interferon-α -stimulated transcription, reconstititued *in vitro*. Genes Dev. 3, 1362-1371.

Li, S., & Sedivy, J. M. (1993). Raf-1 protein kinase activates the NF-κB transcription factor by dissociating the cytoplasmic NF-κB-IκB complex. Proc. Natl. Acad. Sci. USA, 90, 9247-9251.

Li, Y., & Jaiswal. A. K. (1992). Regulation of human NAD(P)H:quinone oxidoreductase gene. J. Biol. Chem. 267, 15097-15104.

Liou, H.-C., & Baltimore, D. (1993). Regulation of the NF-κB/rel transcription factor and IêB inhibitor system. Curr. Opin. Cell Biol. 5, 477-487.

Liu, J., Liu, Y. P., Sendelbach, S. E., & Klaassen, C. D. (1991). Endotoxin induction of hepatic metallothionein is mediated through cytokines. Toxicol. Appl. Pharmacol. 109, 235-240.

Lo, S. K., Janakidevi, K., Lai, L., & Malik, A. B. (1993). Hydrogen peroxide-induced increase in endothelial adhesiveness is dependent on ICAM-1 activation. Am. J. Physiol. 264 (Lung Cell. Mol. Physiol. 8), L406-L412.

Los, M., Dröge, W., Stricker, K., Baeuerle, P. A., & Schulze-Osthoff, K. (1995). Hydrogen peroxide as a potent activator of T lymphocyte functions. Eur. J. Immunol. 25, 159-165.

Mai, S., Stein, B., van den Berg, S., Kaina, B., Lücke-Huhle, C., Ponta H., Rahmsdorf, H. J., Kraemer, M., Gebel, S., & Herrlich, P. (1989). Mechanisms of the ultraviolet light response in mammalian cells. J. Cell. Sci. 94, 609-615.

Marais, R., Wynne, J., & Treisman, R. (1993). The SRF accessory protein Elk-1 contains a growth factor-regulated transcriptional activation domain. Cell 73, 381-393.

Marsh, J. P., & Mossman, B. T. (1991). Role of asbestos and active oxygen species in activation and expression of ornithine decarboxylase in hamster tracheal epithelial cells. Cancer Res. 51, 167-173.

Marui, N., Offermann, M. K., Swerlick, R., Kunsch, C., Rosen, C. A., Ahmad, M., Alexander, R. W., & Medford, R. M. (1993). Vascular cell adhesion molecule-1 (VCAM-1) gene transcription and expression are regulated through an antioxidant-sensitive mechanism in human vascular endothelial cells. J. Clin. Invest. 92, 1866-1874.

Meichle, A., Schütze, S., Hensel, G., Brunsing, D., & Krönke, M. (1990). Protein kinase C-independent activation of nuclear factor κB by tumor necrosis factor. J. Biol. Chem. 265, 8339-8343.

Meier, B., Radeke, H. H., Selle, S., Sies, H., Resch, K., & Habermehl, G. G. (1989). Human fibroblasts release reactive oxygen species in response to interleukin-1 or tumor necrosis factor-α. Biochem. J. 263, 539-545.

Meyer, M., Schreck, R., & Baeuerle, P. A. (1993). H_2O_2 and antioxidants have opposite effects on activation of NF-κB and AP-1 in intact cells: AP-1 as secondary antioxidant-responsive factor. EMBO J. 12, 2005-2015.

Meyer, M., Caselmann, W. H., Schlüter, V., Schreck, R., Hofschneider, P. H., & Baeuerle, P. A. (1992). Hepatitis B virus transactivator MHBs^t: activation of NF-κB, selective inhibition by antioxidants and integral membrane localization. EMBO J. 11, 2991-3001.

Mihm, S., Ennen, J., Pessara, U., Kurth, R., & Dröge, W. (1991). Inhibition of HIV-1 replication and NF-κB activity by cysteine and cysteine derivatives. AIDS 5, 497-503.

Molitor, J. A., Ballard, D. W., & Green, W. C. (1991). κB-specific DNA binding properties are differentially inhibited by enhancer mutations and biological oxidation. New Biol. 3, 987-991.

Müller, J. M., Cahill, M. A., Rupec, R. A., Baeuerle, P. A., & Nordheim, A. (1997). Antioxidants as well as oxidants activate c-fos via Ras-dependent activation of extracellular-signal-regulated kinase 2 and Elk-1. Eur. J. Biochem. 244, 45-52.

Murrell, G. A. C., Francis, M. J. O., & Bromley L. (1990). Modulation of fibroblast proliferation by oxygen free radicals. Biochem. J. 265, 659-665.

Myrset, A. H., Bostad, A., Jamin, N., Lirsac, P.-N., Toma, F., & Gabrielsen, O. S. (1993). DNA and redox state induced conformational changes in the DNA-binding domain of the Myb oncoprotein. EMBO J. 12, 4625-4633.

Nabel, G., & Baltimore, D. (1987). An inducible transcription factor activates expression of human immunodeficiency virus in T cells. Nature 326, 711-713.

Nakamura, K., Hori, T., Sato, N., Sugie, K., Kawakami, T., & Yodoi, J. (1993). Redox regulation of a src family protein tyrosine kinase $p56^{lck}$ in T cells. Oncogene 8, 3133-3139.

Narayanan, R., Klement, J. F., Ruben, S. M., Higgins, K. A., & Rosen, C. A. (1992). Identification of a naturally occuring transforming variant of the p65 subunit of NF-κB. Science 256, 367-370.

Neri, A., Chang, C. C., Lombardi, L., Salina, M., Corradini, P., Maiolo, A. T., Chagnati, R. S., & Dalla, F. R. (1991). B cell lymphoma-associated chromosomal translocation involves canditate oncogene lyt-10, homologous to NF-κB. Cell 67, 1075-1087.

Nolan, G. P., & Baltimore, D. (1992). The inhibitory ankyrin and activator Rel proteins. Curr. Opin. Gen. Dev. 2, 211-220.

Nose, K., Shibanuma, M., Kikuchi, K., Kageyama, H., Sakiyama, S., & Kuroki, T. (1991). Transcriptional activation of early-response genes by hydrogen peroxide in a mouse osteoblastic cell line. Eur. J. Biochem. 201, 99-106.

Ohno, H., Takimoto, G., McKeithan, T. W. (1990). The candidate proto-oncogene bcl-3 is related to genes implicated in cell lineage determination and cell cycle control. Cell 60, 991-997.

Osborn, L., Kunkel, S., & Nabel, G. J. (1989). Tumor necrosis factor α and interleukin 1 stimulate human immunodeficiency virus type 1 through induction of a nuclear factor binding to the NF-κB sites of the long terminal repeat. Proc. Natl. Acad. Sci. USA 86, 5974-5978.

Radler-Pohl, A., Sachsenmaier, C., Gebel, S., Auer, H.-P., Bruder, J. T, Rapp, U., Angel, P., Rahmsdorf, H. J., & Herrlich, P. (1993). UV-induced activation of AP-1 involves obligatory extranuclear steps including Raf-1 kinase. EMBO J. 12, 1005-1012.

Roeder, R. G. (1991). The complexities of eukaryotic transcription initiation: regulation of preinitiation complex assembly. Trends. Biochem. Sci. 16, 402-408.

Roederer, M., Staal, F. J., Ela, S. W., Herzenberg, L. A., & Herzenberg, L. A. (1993). N-acetylcysteine: potential for AIDS therapy. Pharmacology 46, 121-129.

Roederer, M., Staal, F. J. T., Raju, P. A., Ela, S. W., Herzenberg L. A., & Herzenberg, L. A. (1990). Cytokine-stimulated human immunodeficiency virus replication is inhibited by N-acetyl-L-cysteine. Proc. Natl. Acad. Sci. USA 87, 4884-4888.

Rossi, F. (1986). The O_2^-- forming NADPH oxidase from phagocytes: nature, mechanism of activation, and function. Biochim. Biophys. Acta 853, 65-89.

Roth, S., & Dröge, W. (1987). Regulation of T-cell activation and T-cell growth factor (TCFG) production by hydrogen peroxide. Cell. Immunol. 108, 417-421.

Roth, S., & Dröge, W. (1991). Regulation of interleukin 2 production, interleukin 2 mRNA expression and intracellular glutathione levels in ex vivo derived T lymphocytes by lactate. Eur. J. Immunol. 21, 1933-1937.

Rushmore, T. H., Morton, M. R., & Pickett, C. B. (1991). The antioxidant response element. J. Biol. Chem. 266, 11632-11639.

Satriano, J. A., Shuldiner, M., Hora, K., Xing, Y., Shan, Z., & Schlondorff, D. (1993). Oxygen radicals as second messengers for expression of the monocyte chemoattractant protein, JE/MCP-1, and the monocyte colony-stimulating factor, CSF-1, in response to tumor necrosis factor-α and immunoglobulin G. J. Clin. Invest. 92, 1564-1571.

Schenk, H., Klein, M., Erdbrügger, W., Dröge, W., & Schulze-Osthoff, K. (1994). Distinct effects of thioredoxin and other antioxidants on the activation of NF-êB and AP-1. Proc. Natl. Acad. Sci. USA 91, 1672-1676.

Schieven, G. L., Kirihara, J. M., Myers, D. E., Ledbetter, J. A., & Uckun, F. M. (1993). Reactive oxygen intermediates activate NF-κB in a tyrosine kinase-dependent mechanism and in combination with vanadate activate the $p56^{lck}$ and $p59^{fyn}$ tyrosine kinases in lymphocytes. Blood 82, 1212-1220.

Schmitz, M. L., Henkel, T., & Baeuerle, P. A. (1991). Proteins controlling the nuclear uptake of NF-κB, Rel and dorsal. Trends Cell Biol. 1, 130-137.

Schreck, R., Rieber, P., & Baeuerle, P. A. (1991). Reactive oxygen species intermediates as apparently widely used messengers in the activation of the NF-κB transcription factor and HIV-1. EMBO J. 10, 2247-2258.

Schreck, R., Meier, B., Männel, D. N., Dröge, W., & Baeuerle, P. A. (1992a). Dithiocarbamates as potent inhibitors of NF-κB activation in intact cells. J. Exp. Med. 175, 1181-1194.

Schreck, R., Albermann, K., & Baeuerle, P. A. (1992b). Nuclear factor κB: an oxidative stress-responsive transcription factor of eukaryotic cells (a review). Free Rad. Res. Comms. 17, 221-237.

Schulze-Osthoff, K., Bakker, A. C., Vanhaesebroeck, B., Beyaert, R., Jacobs, W., & Fiers, W. (1992). Cytotoxic activity of tumor necrosis factor is mediated by early damage of mitochondrial functions. Evidences for the involvement of mitochondrial radical generation. J. Biol. Chem. 267, 5317-5323.

Schulze-Osthoff, K., Beyaert, R., Vandevoorde, V., Haegeman, G., & Fiers, W. (1993). Depletion of the mitochondrial electron transport abrogates the cytotoxic and gene-inductive effects of TNF. EMBO J. 12, 3095-3104.

Schulze-Osthoff, K., Fiers, W., & Dröge, W. (1994). Endothelial expression of interleukin-6 and ELAM-1 is inhibited by antioxidants and reactive oxygen scavengers. submitted.

Sen, R., & Baltimore, D. (1986). Mutiple nuclear factors interact with the immunoglobulin enhancer sequence. Cell 46, 705-716.

Sherman, M. L., Datta, R., Hallahan, D. E., Weichselbaum, R. R., & Kufe, D. W. (1991). Regulation of tumor necrosis factor gene expression by ionizing radiation in human myeloid leukemia cells and peripheral blood monocytes. J. Clin. Invest. 87, 1794-1797.

Shibanuma, M., Kuroki, T., & Nose, K. (1990). Stimulation by hydrogen peroxide of DNA synthesis, competence family gene expression and phosphorylation of a specific protein in quiescent Balb/3T3 cells. Oncogene 5, 1025-1032.

Shirakawa, F., & Mizel, S. B. (1989). In vitro activation and nuclear translocation of NF-κB catalyzed by cyclic AMP-dependent protein kinase and protein kinase C. Mol. Cell Biol. 9, 2424-2430.

Shreeniwas, R., Koga, S., Karakurum, M., Pinsky, D., Kaiser, E., Brett, J., Wolitzky, B. A., Norton, C., Plocinski, J., Benjamin, W., Burns, K. K., Goldstein, A., & Stern, D. (1992). Hypoxia-mediated induction of endothelial cell interleukin-1α. J. Clin. Invest. 90, 2333-2339.

Staal, F. J. T., Ela, S. W, Roederer, M., Anderson, M. T., Herzenberg, L. A., & Herzenberg, L. A. (1992). Glutathione deficiency and human immunodeficiency virus infection. Lancet 339, 909-912.

Staal, F. J. T., Roederer, M., Herzenberg, L. A., & Herzenberg, L. A. (1990). Intracellular thiols regulate activation of nuclear factor κB and transcription of human immunodeficiency virus. Proc. Natl. Acad. Sci. USA 87, 9943-9947.

Stein, B., Rahmsdorf, H. J., Steffen, A., Litfin, M., & Herrlich, P. (1989). UV-induced DNA damage is an intermediate step in UV-induced expression of human immunodeficiency virus type 1, collagenase, *c-fos*, and metallothionein. Mol. Cell. Biol. 9, 5169-5181.

Storz, G., Tartaglia, L. A., Farr, S. B., & Ames, B. N. (1990a). Bacterial defenses against oxidative stress. Trends Genet. 6, 363-368.

Storz, G., Tartaglia, L. A., & Ames, B. N. (1990b). Transcriptional regulator of oxidative stress-inducible genes: direct activation by oxidation. Science 248, 189-194.

Sun, S.-C., Ganchi, P. A., Ballard, D. W., & Greene, W. C., (1993). NF-κB controls expression of inhibitor IκB-α: Evidence for an inducible autoregulatory pathway. Science 259, 1912-1915.

Suzuki, Y. J., & Packer, L. (1993). Inhibition of NF-κB activation by vitamin E derivatives. Biochem. Biophys. Res. Commun. 193, 277-283.

Suzuki, Y., Wang, W., Vu, T. H., & Raffin, T. A. (1992). Effect of NADPH oxidase inhibiton on endothelial ELAM-1 mRNA expression. Biochem. Biophys. Res. Commun. 184, 1339-1343.

Toledano, M. B., & Leonard, W. J. (1991). Modulation of transcription factor NF-κB binding by oxidation-reduction *in vitro*. Proc. Natl. Acad. Sci. USA 88, 4328-4332.

Treisman, R. (1992). The serum response element. Trends Biochem. Sci. 17, 423-426.

Trush, M. A., & Kensler, T. W. (1991). An overview of the relationship between oxidative stress and chemical carcinogenesis. Free Rad. Biol. Med. 10, 201-210.

Turrens, J. F, Alexandre, A., & Lehninger, A. L. (1985). Ubisemiquinone is the electron donor for superoxide formation by complex III of heart mitochondria. Arch. Biochem. Biophys. 237, 408-414.

Vassalli, P. (1992). The pathophysiology of tumor necrosis factor. Ann. Rev. Immunol. 10, 411-452.

Visner, G. A., Dougall, W. C., Wilson, J. M., Burr, I. A., & Nick, H. S. (1990). Regulation of manganese superoxide dismutase by lipopolysaccharide, interleukin-1, and tumor necrosis factor. J. Biol. Chem. 265, 2856-2864.

Wasylyk, C., & Wasylyk, B. (1993). Oncogenic conversion of Ets affects redox regulation *in-vivo* and *in-vitro*. Nucleic Acids Res. 21, 523-529.

White, C. W., Ghezzi, P., Dinarello, C. A., Caldwell, S. A., McMurthy, I. F., & Repine, J. E. (1987). Recombinant tumor necrosis factor/cachectin and interleukin 1 pretreatment decreases lung oxidized glutathione accumulation, lung injury, and mortality in rats exposed to hyperoxia. J. Clin. Invest. 79, 1868-1873.

White, C. W., Ghezzi, P., McMahon, S., Dinarello, C. A., & Repine, J. E., (1989). Cytokines increase lung antioxidant enzymes during exposure to hyperoxia. J. Appl. Physiol. 66, 1003-1007.

Witte, L., Fuks, Z., Haimonovitz-Friedman, H., Vlodavsky, I., Goodman, D. S., & Eldor, A. (1989). Effects of irradiation on the release of growth factors from cultured bovine, porcine, and human endothelial cells. Cancer Res. 49, 5066-5072.

Wong, G. H. W., Elwell J. H., Oberley, L. W., & Goeddel, D. V. (1989). Manganeous superoxide dismutase is essential for cellular resistance to cytotoxicity of tumor necrosis factor. Cell 58, 923-931.

Wong, G. H. W., & Goeddel, D. V. (1988). Induction of manganeous superoxide dismutase by tumor necrosis factor: possible protective mechanisms. Science 242, 941-944.

Wu, J., & Weiss, B. (1991). Two divergently transcribed genes, *soxR* and *soxS*, control a superoxide response regulon in *Escherichia coli*. J. Bacteriol. 173, 2864-2871.

Xanthoudakis, S., Miao, G., Wang, F., Pan, Y.-C. E., & Curran, T. (1992). Redox activation of Fos-Jun DNA-binding activity is mediated by a DNA repair enzyme. EMBO J. 11, 3323-3335.

Yamauchi, N., Kuriyama, Y. H., Watanabe, N., Neda, H., Maeda, M., & Niitsu, Y. (1989). Intracellular hydroxyl radical production induced by recombinant human tumor necrosis factor and its implication in the killing of tumor cells *in vitro*. Cancer Res. 49, 1671-1675.

Yodoi, J., & Uchiyama, T. (1992). Diseases associated with HTLV-I virus, IL-2 receptor dysregulataion and redox regulation. Immunol. Today 13, 405-411.

Zabel, U., & Baeuerle, P. A. (1990) Purified human IκB can rapidly dissociate the complex of the NF-κB transcription factor with its cognate DNA. Cell 61, 255-265.

Zabel, U., Henkel, T., dos Santos Silva, M., & Baeuerle, P. A. (1993). Nuclear uptake control of NF-κB by MAD-3, an IκB protein present in the nucleus. EMBO J. 12, 201-211.

Ziegler, D. M. (1985). Role of reversible oxidation-reduction of enzyme thiol-disulfides in metabolic regulation. Ann. Rev. Biochem. 54, 305-329.

Zhu, H., Roy, A. L., Roeder, R. G., & Prywes, R., (1991). Serum response factor affects preinitiation complex formation by TFIID. New Biologist 3, 455-464.

Zimmerman, R. J., Chan, A., & Leadon, S. A. (1989). Oxidative damage in murine tumor cells treated *in vitro* by recombinant human tumor necrosis factor. Cancer Res. 49, 1644-1648.

Zinck, R., Hipskind, R. A., Pingoud, V., & Nordheim, A. (1993). *c-fos* transcriptional activation and repression correlate temporally with the phosphorylation status of TCF. EMBO J. 12, 2377-2387.

OXYRADICALS AND MALIGNANT TRANSFORMATION

Larry W. Oberley and Terry D. Oberley

I. INTRODUCTION

Over the last two decades a great deal of evidence has accumulated linking reactive oxygen species (ROS) and cancer. ROS are molecules that contain oxygen and

Advances in Molecular and Cellular Biology
Volume 25, pages 45-69.

ISBN: 0-7623-0379-4

have higher reactivity than ground state molecular oxygen. These species include not only the oxygen radicals (such as superoxide, hydroxyl, and peroxyl radicals), but also non-radical molecules such as singlet oxygen and hydrogen peroxide. ROS are generated during normal aerobic metabolism and increased levels of these species are produced during various forms of oxidative stress. ROS are known to react with many intracellular targets, including lipids, proteins, and DNA. ROS-induced damage can result in cell death, mutations, chromosomal aberrations, or carcinogenesis (Cerutti, 1985). The intracellular concentration of ROS is the consequence of the production of ROS and the ability of substances to remove them.

Cells contain a large number of antioxidants to prevent or repair the damage caused by ROS. These include a number of small molecular weight antioxidants such as vitamins C, A, and E, as well as the larger molecular weight antioxidant enzymes. There are four known primary intracellular antioxidant enzymes in mammalian cells—copper and zinc containing superoxide dismutase (CuZnSOD), manganese containing superoxide dismutase (MnSOD), catalase (CAT) and peroxidase, of which glutathione peroxidase (GPX) appears to be the most important. The SODs convert superoxide radical into hydrogen peroxide, while the catalases and peroxidases convert hydrogen peroxide into water. In this way, two toxic species, superoxide radical and hydrogen peroxide, are converted to the harmless product water. These enzymatic functions are thought to be necessary for life in all oxygen metabolizing cells (McCord et al., 1971). SOD and CAT need no co-participants to function, but GPX requires the presence of reduced glutathione and secondary enzymes for replenishing this reductant. The antioxidant enzyme scheme is shown in Figure 1 where GSH is reduced glutathione and

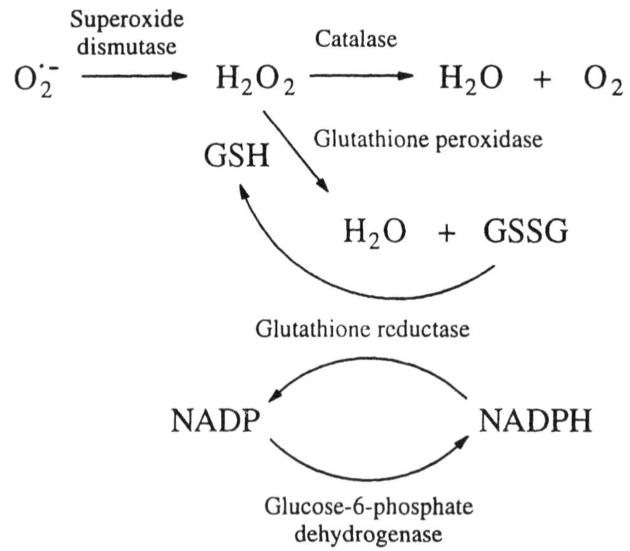

Figure 1. Relationships among the antioxidant enzymes.

GSSG is oxidized glutathione. In this scheme, glucose-6-phosphate dehydrogenase and glutathione reductase are considered secondary antioxidant enzymes since they do not act on ROS, but they enable GPX to function. In addition, an enzyme not shown on this diagram, glutathione S-transferase, can be considered a primary antioxidant enzyme because it removes hydroperoxides.

An important feature of these enzymes is that they are highly compartmentalized. MnSOD is found in the mitochondria, CuZnSOD in the cytoplasm, CAT in peroxisomes and cytoplasm, and GPX in many subcellular compartments. There is also an extracellular SOD (ECSOD) which contains Cu and Zn; this enzyme is secreted into extracellular spaces. One reason for the existence of the many forms of these enzymes is to reduce oxidative stress in the various intracellular and extracellular compartments.

II. THE ROLE OF ROS IN CHEMICAL CARCINOGENESIS

Chemical carcinogenesis is thought to occur in two distinct steps: initiation and promotion. Recent literature also usually includes a third step: progression. Pure initiators cannot cause tumor formation by themselves and thus are termed incomplete carcinogens. Most compounds that are classified as initiators cause cancer by themselves (i.e., are complete carcinogens) because they demonstrate both initiating and promoting activity. Promoters show little or no initiating action and thus induce few tumors by themselves: they are also incomplete carcinogens. Progressors are substances that complete the malignant transformation process by converting benign tumors to malignant tumors; progression is usually effected by reapplication of the initiator some time after the initial application.

Initiators are in general metabolized in cells to ultimate carcinogens, which are thought to act by forming DNA adducts. Initiators cause mutations. ROS have also been demonstrated to cause mutation (for reviews see Oberley, 1982, and Oberley and Oberley, 1986). As examples, two compounds which produce superoxide radical, potassium superoxide (Cunningham and Lokesh, 1983) and bleomycin (Cunningham et al. 1984), cause mutation in mammalian cells that is blocked by SOD. Initiators have also been shown to cause cellular ROS production (Oberley, 1982; Oberley and Oberley, 1986).

In contrast to initiators, tumor promoters bind to the plasma membrane; tumor promoters do not bind to DNA or cause mutation, but promoters cause DNA strand breaks (Birnboim, 1981). Promoters also cause the generation of ROS (for reviews, see Kensler and Trush, 1984, and Oberley and Oberley, 1986). The production of ROS by tumor promoters has been demonstrated to be due to two cellular sources: white blood cells (Goldstein et al., 1981) and target cells of the carcinogenic process (Miller et al., 1982). The production of ROS by white blood cells cannot be the sole causative agent in carcinogenesis since cells in culture can be malignantly transformed when no white cells are present.

The role for ROS in malignant transformation is strengthened by two observations: (1) ROS themselves can act as initiators (Cunningham and Lokesh, 1983; Cunningham et al., 1984) and promoters of carcinogenesis (Zimmerman and Cerutti, 1984), and (2) scavengers of ROS can inhibit mutation, chromosomal aberration, initiation, promotion, and complete malignant transformation (Oberley and Oberley, 1986). Scavengers of ROS have been demonstrated to inhibit these processes both *in vitro* and *in vivo*. Scavengers which have been shown to inhibit in various systems include SOD, CAT, mannitol, benzoate, sulfhydryl compounds, vitamin E, and synthetic SODs.

The observation that SOD and CAT when added to culture medium had dramatic effects on DNA damage and transformation suggests an extracellular site of action since these enzymes do not penetrate well into cells. This was most confusing initially since intracellular DNA is thought to be the target of carcinogenesis. Cerutti et al. (1983) have proposed a theory that circumvents this problem. They hypothesized that tumor promoters bind to membranes, stimulate the arachidonic acid cascade, and disturb the structural integrity of the membranes. Lipid hydroperoxides and aldehydic components were suggested as mediators between the membrane and the genome. Lipid hydroperoxides were proposed to be active oxygen carriers and stabilizers, which eventually release ROS and degrade to aldehydic compounds. Lipid hydroperoxides and their degradation products were proposed to produce chromosomal damage not only in the stimulated cells but also in neighboring tissue; i.e., they are clastogenic (chromosome-breaking) factors.

Taken together, the data indicate that chemical carcinogens produce ROS and that these ROS are responsible for a certain fraction of cancer incidence. These conclusions are strengthened by a recent paper that used a molecular biological approach. St. Clair et al. (1992) showed that overexpression of MnSOD via cDNA transfection led to suppression of ionizing radiation-induced, but not chemically-induced transformation. Transfection of the human MnSOD cDNA into mouse C3H 10T1/2 cells resulted in the production of active human MnSOD, that was properly transported into the mitochondrial matrix. The clone overexpressing MnSOD was about twice as resistant to radiation-induced malignant transformation as the control cells. In contrast, there was no difference in transformation frequency in cells exposed to the chemical carcinogen, 3-methyl-cholanthrene. These data show for the first time that elevation of endogenous antioxidant enzyme leads to suppression of malignant transformation and suggests that a similar molecular biological approach might be successful in preventing malignancy in families with genetic disposition to cancer or in groups with high exposure to certain agents that cause cancer (i.e., radiation).

III. TUMOR CELL ANTIOXIDANT ENZYME LEVELS

The above work shows quite clearly that carcinogenesis causes the cellular production of ROS. It has been demonstrated in normal cells that in general elevated generation of ROS will lead to induction of one or more of the antioxidant enzymes. However, it has also been shown that the final product of the carcinogenic process and its associated oxidative stress—the tumor cell—has in general lowered antioxidant enzymes. It has been found that tumor cells are nearly always low in MnSOD and CAT and usually low in CuZnSOD activity (Oberley and Buettner, 1979; Oberley, 1982; Oberley and Oberley, 1986). GPX activity shows great variablility. One would expect just the opposite: tumor cells should have elevated antioxidant enzymes to meet the oxidative insults demonstrated in carcinogenesis.

Before examining the possible reasons for this paradox, we will review the literature on the levels of antioxidant enzymes in tumor cells. We will focus on only one of these enzymes—MnSOD—since this is the one on which most work has been done. It should be emphasized that the other antioxidant enzymes are also undoubtedly important in carcinogenesis. A summary of the knowledge of MnSOD gathered to date is listed below:

1. MnSOD activity is diminished in tumor cells only when compared to a proper cell type—the normal cell type from which the tumor arose (Oberley and Oberley, 1986). In other words, tumors do not always demonstrate a near zero MnSOD activity, but rather tumor MnSOD activity is lowered compared to a proper control. Comparisons to whole organs are usually not valid, since most organs are comprised of many cell types. Tumors are clonal in nature and thus arise from only one cell type; it is to this one cell type that the tumor must be compared.
2. The reduction in tumor MnSOD activity has been observed in all species examined and does not depend on the transforming agent (Oberley and Buettner, 1979; Oberley, 1982; Oberley and Oberley, 1986). Thus, lowered MnSOD activity has been observed in human and various rodent (rat, mouse, hamster) cells, whether spontaneously transformed (Sun et al. 1993a) or transformed by viruses, chemicals, ionizing radiation, or hormones (McCormick et al., 1991).
3. In the tumor cells examined thus far, MnSOD activity was diminished because the levels of MnSOD mRNA and immunoreactive protein were lowered when compared to an appropriate normal cell control (McCormick et al.,1991; Oberley et al., 1989a; Marlens et al., 1985; Galeotti et al., 1989; St.Clair and Oberley, 1991; Sun et al., 1993b). At least part of the reason that tumor cells are low in MnSOD activity is because the amount of translatable mRNA is low and thus less protein is synthesized. In the tumors examined thus far, the coding region of the MnSOD gene was normal, suggesting changes in regulatory regions of the MnSOD gene or changes in regulatory protein(s) that bind to these regulatory regions (St. Clair and Holland, 1991).
4. Inducible enzyme levels may be diminished in cells because their substrate levels are low and unable to induce biosyntheis of that enzyme. However, several studies have demonstrated that cancer cells have the capacity to produce superoxide radicals ($O_2^{\cdot-}$), the substrate for SOD (Oberley and Buettner, 1979; Bize et al., 1980; Oberley, 1982; Konstantinov et al., 1987). Thus, it appears that the reason

why SOD, an inducible enzyme, is diminished in tumor cells is not due to a low production of metabolically produced superoxide radicals. This suggests that superoxide generation coupled with diminished amounts of MnSOD may be a general characteristic of tumor cells. It has also been hypothesized that tumor cells have diminished MnSOD activities because they have lowered intracellular oxygen concentrations. *In vivo* tumors are often characterized by a hypoxic core. Indeed, it has been observed that the total levels of SOD are lowered in the hypoxic centers of certain animal tumors (Petkau et al., 1977). However, this cannot be the cause for the general lowering of tumor cell MnSOD since diminished levels of MnSOD have consistently been observed in transformed tissue culture cells compared to normal cells: in this case the levels of oxygen are equivalent in the normal and tumor cells.

5. Not only are the constitutive levels of MnSOD low in tumor cells, but tumor cells have also lost much of the ability to undergo MnSOD induction upon exposure to oxygen or superoxide radicals (Oberley and Oberley, 1986; Oberley et al., 1987). Until recently, oxidative stress was the only known way to cause MnSOD induction. It has now been demonstrated that tumor necrosis factor (TNF) can cause MnSOD induction in some, but not all tumor cells (Wong and Goeddel, 1988; Wong et al., 1989; Asoh et al., 1989; Kawaguchi et al., 1990, Warner et al., 1991; Pang et al., 1992; Himeno et al., 1992). Recently, other cytokines (Wong and Goeddel, 1988; Ono et al., 1992; Marklund, 1992; Masuda et al., 1988) and the cytokine mimic phorbol ester (Fujii and Taniguchi, 1991) have also been shown to induce MnSOD in some tumor cells. The reason why oxidative stress fails to induce MnSOD in tumor cells while cytokines are successful is still unknown.

Most of the above data on induction of MnSOD by cytokines rely solely on the measurement of mRNA levels. Recently, we have clearly demonstrated several examples of the induction of MnSOD mRNA, but not a concurrent increase in MnSOD protein or activity (Czaja et al., 1994). This means that most of the above data need to be reexamined and it is possible that true MnSOD induction may not occur in some of the tumor lines examined.

6. We and others have hypothesized that malignant transformation requires at least two steps: immortalization and loss of control of cell proliferation (Oberley and Oberley, 1986). Correlative evidence suggests that the changes in MnSOD observed in tumor cells is due to immortalization and not due to defects in the control of cell proliferation (Oberley et al., 1978; Oberley et al., 1989b; Loven et al., 1984). Several groups are currently examining this question using MnSOD sense cDNA transfection into immortal, but non-malignant cell lines (cells that have indefinite growth potential, but do not form tumors in syngeneic hosts or nude mice).

Evidence also suggests that the second step of malignant transformation—loss of control of cell proliferation—also involves ROS and CuZnSOD (and possibly CAT) (Armato et al., 1984; Oberley, 1985; Craven et al., 1986; Shibanuma et al., 1988; Burdon and Rice-Evans, 1989; Murrell et al., 1990). The most convincing evidence involved studying the effect of liposomal CuZnSOD or CAT on diethylstilbestrol (DES)-induced cell proliferation (Oberley et al., 1991). DES initiated proliferation of Syrian hamster renal proximal tubular cells. Liposomes containing CuZnSOD or CAT suppressed DES-induced proliferation, whereas empty liposomes or liposomes containing inactivated CuZnSOD did not. Liposomes containing CuZnSOD or CAT did not inhibit cell proliferation in the absence of DES. In the absence of liposomes, exogenous CuZnSOD did not suppress DES-induced proliferation. The decrease in cell numbers observed in CuZnSOD or CAT-containing liposomes was not due to decreased cell viability. These results demonstrate that ROS and CuZnSOD and CAT play a role in cell division.

7. In both normal and tumor cells, activities of SOD, particularly MnSOD, correlate with the degree of differentiation of the cells (Bize et al., 1980; Oberley and Oberley, 1986). In general, the more differentiated the cell, the higher the SOD activity. This conclusion has been based on tissue culture work and whole tissue homogenates from animals. The latter studies do not take into account that most tissues are comprised of many cell types. We have recently provided strong evidence *in vivo* for the relationship between SOD and differentiation using a technique which overcomes the heterogeneity problem of tissues: immunohistochemistry allows us to examine all cell types and demonstrated strong induction of MnSOD, CuZnSOD, and CAT during differentiation of hamster tissue (Oberley et al., 1990). Induction of these enzymes occurred not only during fetal development, but also in the adult during cell renewal and subsequent differentiation. Thus, staining of these three proteins occurred in differentiated epithelial cells, but not in epithelial stem cells. We have hypothesized that tumor cells have low MnSOD because they are derived from stem or transit cells that have low MnSOD (Oberley and Oberley, 1986). Another piece of information linking SOD to differentiation is the observation that liposomal CuZnSOD induced differentiation in a mutant non- differentiating strain of the slime mold *Physarum polycephalum* (Allen et al., 1988); the authors of this paper concluded that "increased SOD activity acts as a physiological stimulus of differentiation" in this species.

Very recently even more powerful evidence for the role of MnSOD in differentiation has been reported. St.Clair et al. (1994) demonstrated that elevation of MnSOD via transfection led to greatly increased levels of differentiation in cells exposed to the demethylating agent 5-azacytidine. C3H 10T1/2 cells exposed to 5-azacytidine alone showed no differentiated muscle cells, while a small fraction of the cells overexpressing MnSOD exhibited spontaneous myoblast differentiation. Three mM 5-azacytidine induced a low frequency of myoblasts in parental cells, but induced extensive formation of myoblasts in MnSOD overexpressing cells. Thus, elevation of MnSOD enormously potentiated the differentiating effect of 5-azacytidine.

8. Loss of MnSOD shows a strong correlation with mitochondrial damage. In one study, mitochondria were isolated from the Morris hepatomas; these hepatomas are a series of stable transplantable rat tumors that have a range of phenotypes from fast-growing, undifferentiated to slow-growing, differentiated tumors (Bize et al., 1980). When the mitochondria from the fast growth rate and medium

growth rate tumors were isolated and then subsequently examined using electron microscopy, it was found that these mitochondria, which had much lower levels of MnSOD than mitochondria from normal rat liver, were severely damaged. In contrast, mitochondria from the slow growth rate tumor had above normal levels of MnSOD and appeared normal morphologically. In this experiment, the mitochondria might be damaged *in situ*, or might be damaged during the isolation procedure. In this case, the organelles would simply be more susceptible to damage. In general, mitochondria in tumors have been found to be abnormal even before isolation. For example, Springer (1980) has characterized the intact mitochondria of 16 human epithelial lines. Hypertrophied mitochondria and longitudinal cristal arrangement were found in almost all the malignant lines, but not in any lines derived from nonmalignant tissues of cancerous organs or from normal tissue.

Abnormal mitochondria are found not only in malignant cells, but also in nonmalignant, immortal, nondifferentiating cells. This was found quite unexpectedly in our study of X-REF-23 rat embryo fibroblast cells (Oberley et al., 1989b). X-REF-23 is an immortal cell line that maintains a nontransformed phenotype throughout its measured lifespan. Low-passage X-REF-23 cells undergo spontaneous differentiation into muscle and adipose cells after reaching confluence, while high-passage X-REF-23 cells undergo little or no differentiation. SOD activities were measured in subclones of X-REF-23, which differentiate into muscle or adipose cells, as well as parental nondifferentiating cells. MnSOD activity was induced during the time period when differentiation occurred in the two differentiating lines. In contrast, MnSOD activity did not increase in this time period in the nondifferentiating cell line. However, MnSOD activity was induced in the latter cell line at a much later time, still without differentiation occurring. This later induction appeared to be caused by a large increase (about 20-fold) in the number of mitochondria. These mitochondria were very small and showed evidence of disorganization. These data indicate that in the nondifferentiating cell line, normal MnSOD induction (via increasing the amount of MnSOD per mitochondria) did not occur. Rather, abnormal MnSOD induction was demonstrated where the number of mitochondria increased. This process somehow led to mitochondrial damage. Thus, it appears that abnormal mitochondria are associated with a lack of differentiation in an immortal cell line. In both tumor cells and nonmalignant, immortal cells, abnormal mitochondria are found in cells with diminished MnSOD activity; these results imply that loss of MnSOD has detrimental structural and possibly functional consequences to mitochondria.

9. An often-quoted scientific saying is that "exceptions prove the rule." This is also true for the generality of the diminished MnSOD found in tumor cells. There have been several reports of tumor cells that did not have lowered MnSOD levels. However, all of these observations were flawed in some way: either they did not have a proper control, or they measured only total SOD and not MnSOD, or they measured mRNA or immunoreactive protein levels and not activity values. We have recently very carefully documented the first known exception to the

rule that diminished MnSOD is found in tumor cells (Oberley et al., 1994). MnSOD activity was measured in a large number of human renal adenocarcinomas and compared to either normal human kidney or isolated human kidney proximal tubules (the cell of origin). The tumor MnSOD activities ranged from much lower than normal to much higher than normal; some of the adenocarcinomas had normal MnSOD activity. In order to understand this unusual variability in renal tumor MnSOD activities, we examined the tissues using immunohistochemical analysis at the light and electron microscopic level. Histological analysis using hematoxylin and eosin staining demonstrated that the renal cell carcinomas could be subclassified into clear cell, granular cell, or mixed clear and granular cell variants; this is the usual subclassification seen in these tumors. In all three types of tumors, immunogold studies showed little staining using antibodies to CuZnSOD and GPX, while the normal proximal tubules demonstrated strong staining for MnSOD, CuZnSOD, CAT, and GPX. Intensity of labeling for MnSOD and CAT depended on the tumor type. Clear cell variants demonstrated trace staining for MnSOD and CAT, while granular cell variants exhibited heavy staining for both enzymes. Mixed types of tumors showed clear cells with trace staining for all antioxidant enzymes, while granular cells again showed intense labeling for MnSOD and CAT. These data show that the variability in activity in human renal carcinoma is due to heterogeneity of cell types in these tumors. Those tumors with normal MnSOD levels occur simply because they consist of two cell types, one with very low MnSOD activity and one with very high MnSOD activity; the observed MnSOD activity is simply an average of these two cell types and this value turns out to be equal to the MnSOD activity of normal kidney or isolated proximal tubules, which have similar MnSOD activities.

We further studied these tumors at the electron microscope level. Using normal kidney proximal tubule as a comparison, immunogold ultrastructural analysis using antibody to MnSOD demonstrated infrequent small lightly-labeled mitochondria in clear cell variants, while granular cell variants exhibited numerous medium-sized heavily-labeled mitochondria. These observations suggest that MnSOD immunoreactive protein was elevated in granular cells both because there was an increase in the number of mitochondria and because the labeling density in mitochondria was increased compared to mitochondria in clear cell types or in normal proximal tubular cells.

Thus, the granular variant of human renal adenocarcinoma represent an exception to the rule that MnSOD is diminished in tumor cells when compared to an appropriate normal cell control. However, the exception proves the rule. In this case, the granular cell variant still has grossly abnormal MnSOD levels; the MnSOD levels are greatly increased compared to normal. Whether MnSOD is increased or decreased, one still has an abnormal antioxidant enzyme level. Too much SOD has been shown to be detrimental to cells as well as too little SOD (Elroy-Stein et al., 1986). It is our hypothesis that elevation of MnSOD in this granular cell variant leads to an oxidant/antioxidant imbalance and is responsible for part of the granular cell phenotype. It is interesting that the proliferation of mitochondria described above for the immortal X-REF-23 cell line is also seen in the granular renal cell carcinoma. In both cases, one observes an elevation of MnSOD associated with an increase in the number of mitochondria per cell. The cause of the mitochondrial proliferation appears to be linked with the increase in MnSOD; the cause of this linkage is an interesting and important area to study in the future.

IV. THE IMPORTANCE OF THE DIMINISHED MnSOD IN TUMOR CELLS

All of the above evidence is circumstantial and does not demonstrate a causal relationship between MnSOD and cancer. As discussed earlier, a primary argument for a causal relationship is that antioxidants in general, and SOD and SOD mimetics in particular, inhibit malignant transformation (Cerutti, 1985; Oberley and Oberley, 1986). Recently, even more powerful evidence has been obtained. Three different techniques have been used to elevate SOD in tumor cells and the effect on the malignant phenotype then determined. These three techniques were: (1) elevation by exposure to a superoxide generator and subsequent isolation of resistant cells, (2) addition of liposomal SOD, and (3) elevation of MnSOD by sense cDNA transfection. Each of these techniques will be be discussed later.

Fernandez-Pol et al. (1982) have shown using drug resistance experiments a relationship between SOD and the malignant phenotype. They examined the effect of paraquat, a known superoxide producer, on Kirsten virus-transformed NRK (normal rat kidney) cells. Kirsten virus-transformed cells had much lower SOD activities than normal NRK cells. Virus-transformed cells were largely killed by paraquat, but a small fraction of these cells became paraquat-resistant. These resistant cells had very high SOD activities. Moreover, the resistant (revertant) cells had an apparently normal cell phenotype: they appeared normal morphologically, did not grow in soft agar, and had a normal saturation density and serum requirement. The authors concluded: "The overall mechanism underlying the reversion of the clone RE8G3 remains to be determined rigorously; however, it may be solely due to the increase in SOD levels, which may be sufficient to result in reversion of all the transformed properties examined." The one weakness of this study is that the authors did not determine which form of SOD was involved; they just measured total SOD activity. Thus, these experiments need to be duplicated and all forms of SOD quantitated.

A second technique to elevate SOD in tumor cells is the use of liposomal SOD. Native SOD does not penetrate well into cells and so liposomes can be used to deliver the protein. Beckman et al. (1989) have shown that Friend erythroleukemia cells differentiate and stop proliferating in the presence of liposomal SOD. Liposomes without SOD caused no differentiation. The authors concluded; "Of central importance is the results of this study support Oberley's hypothesis that SOD can induce differentiation in tumor cells. Implicit in this hypothesis is the tenet that loss of SOD activity may be in some way causal to cellular transformation." It should be emphasized that this study used CuZnSOD. No one has yet attempted the analogous experiment with liposomal MnSOD. It is possible that both forms of SOD suppress the malignant phenotype.

A third way to deliver SOD to tumor cells is the use of cDNA transfection. In collaboration with Drs. Sue Church and James Grant at Washington University, we have shown that transfection of MnSOD cDNA into cultured human melanoma cells resulted in the loss of the malignant phenotype (1993). The malignant phenotype was tested both *in vitro* by assays such as growth in soft agar and, more importantly, *in vivo* by growth in nude mice. All of these tests showed a loss of the malignant phenotype in clones which overexpressed MnSOD by at least five-fold. The most important observation was that in the nude mouse assay, 18 out of 18 sites injected with the parental melanoma cell line developed tumors, while 0 out of 16 sites injected with melanoma cells containing high levels of MnSOD developed tumors. Interestingly, clones which only had a three-fold elevation of MnSOD formed tumors in nude mice, suggesting a threshold level of the enzyme was necessary to suppress malignancy.

Thus, all three methods to increase SOD in tumor cells have led to the same conclusion: elevation of SOD leads to suppression of the malignant phenotype. However, in science one can never prove hypotheses, but only disprove them. Hence, the meaning of these experiments is still open to debate. A central question is why elevation of SOD has any effect. Tumors could fail to form *in vivo* for a number of reasons: direct killing of the cells because of SOD elevation, differentiation and subsequent death of the tumor cells, apoptosis, or a heightened immune attack. Histological studies are needed to determine the cause of the lack of tumor growth. Another important question that needs to be answered is whether anything else changes after elevation of SOD. Several studies have reported that elevation of CuZnSOD after transfection leads to cells with high GPX (Ceballos et al., 1988; Kelner and Bagnell, 1990). To explain this increased GPX activity, it has been repeatedly proposed that elevation of SOD leads to increased hydrogen peroxide levels, since peroxide is the product of superoxide dismutation (Ceballos, 1988; Kelner and Bagnell, 1990). A direct demonstration of peroxide elevation has not been attempted. In the same vein, hydrogen peroxide is known to induce many proteins. Does this induction occur in cells with elevated SOD levels? If so, are any of the proteins inducible by peroxide actually responsible for the malignant reversion caused by SOD elevation? Finally, since changes in SOD levels affect redox potential and redox potential is known to affect signal transduction pathways (see below), it is possible that increased SOD levels inhibit cellular proliferation by affecting redox-regulated molecules in the signal transduction pathways. These questions are of utmost importance if we are to understand the role of SOD in cancer.

V. CAUSE OF SOD LOSS IN TUMOR CELLS

The above work shows an intimate relationship between the loss of SOD, particularly MnSOD, and the cancer cell phenotype. What is responsible for the diminished MnSOD observed in cancer cells? Two hypotheses have been put forth. As mentioned earlier, one hypothesis is that tumor cells are low in SOD because they are stem or transit cells; stem and transit cells from normal tissue have been demonstrated to be low in SOD activity (Oberley et al., 1990).

A second hypothesis that has been proposed is that MnSOD is low in tumor cells because oxidative stress has caused damage to the gene for this protein (Oberley and Oberley, 1984). The concept is that during carcinogenesis ROS interact with the promoter of certain genes and cause damage so that MnSOD cannot be fully induced. Damage could be to the regulatory regions of the MnSOD gene itself or to other genes which regulate MnSOD expression. No evidence has yet been reported for damage to the coding region of the MnSOD gene, but this could be because so few tumors have been examined at the DNA level. Since the carcinogenic process is somewhat random, it is reasonable to expect many forms of damage, including some to the structural gene.

These two hypotheses can be integrated into the following model. During carcinogenesis, ROS are generated in cells. The effect of these ROS will depend on the cell type. In differentiated cells, generation of increased levels of ROS will lead to induction of the antioxidant enzymes and if this induction is not adequate to protect the cell, death will occur. In fact, induction of antioxidant enzymes early in carcinogenesis has been reported in two model systems (Sun et al., 1988; McCormick et al., 1991). In stem or transit cells, generation of ROS leads to cell proliferation and cell differentiation. In a small fraction of these stem and transit cells, ROS generation will lead to damage to the MnSOD gene or genes that control its induction during differentiation. This damage will lead to proliferation of a stem or transit cell which cannot undergo full MnSOD induction and also cannot terminally differentiate. Thus, in these cells, proliferation and differentiation are no longer coupled as they are in most normal cells. This model explains the paradox described earlier: oxidative stress induces antioxidant enzymes in normal cells, but the final product of the carcinogenic process is a cell which has diminished antioxidant enzymes. Thus, it appears that during carcinogensis, ROS are produced which cause induction of certain of the antioxidant enzymes. However, the stem or transit cells are caused to proliferate and these cells have low antioxidant enzymes. They replace the normal cells and as explained above have lost the capacity to induce these enzymes, particularly MnSOD. The final result is a tumor cell with diminished levels of the antioxidant enzymes.

This model implies that MnSOD induction is necessary for terminal differentiation to occur in most cells. We hypothesize that MnSOD induction is necessary but not sufficient for most cells to differentiate. This means that other steps besides MnSOD induction, like induction of ROS, are also necessary for terminal differentiation to occur. Conversely, MnSOD induction is necessary so that even if all the other steps have occurred, no differentiation can take place unless MnSOD induction has occurred. We envision that differentiation involves the synthesis and maintenance of specialized structures which characterize that special type of cell. These specialized structures undergo large amounts of oxidative metabolism during their function. Thus, they produce high levels of ROS. These ROS must be scavenged by antioxidant enzymes if the structures are to be maintained. A particular structure of importance in this process are the mitochondria and this is why MnSOD plays a central role. Other antioxidant enzymes may be of crucial importance in maintaining the specialized structures of other differentiated cells.

VI. THE ROLE OF ROS IN APOPTOSIS

The above model of carcinogenesis focuses on the stem cell and cellular proliferation. Recent work has suggested that inhibition of cell death can also cause cancer (Williams, 1991). Cell death can be caused by necrosis, or by apoptosis. Apoptosis is programmed cell death. General characteristics of apoptosis include nuclear condensation and degradation of DNA into oligonucleosomal fragments. Inhibition of apoptosis is thought to cause certain cancers because if the time to cell death is prolonged, cell numbers will increase. In other words, in a tumor cell population, the number of tumor cells will depend on both the birth rate and death rate of the population. If the birth rate stays constant, while the death rate decreases, the population numbers will increase. Thus, inhibition of apoptosis will lead to growth of a tumor, as will increased cell proliferation.

Pierce et al. (1991) first proposed that hydrogen peroxide was a mediator of programmed cell death. Previous work had identified in blastocele fluid a soluble factor which killed embryonal carcinoma cells with trophectodermal potential, but not those with embryonic potential (Parchment et al., 1990). This toxic activity was postulated to be hydrogen peroxide, generated from amine oxidases (Gramzinski et al., 1990). Pierce et al. (1991) observed that blastocele fluid was not toxic to malignant pretrophectodermal cells in the presence of CAT, but was toxic in its absence. Malignant cells with embryonic potential that normally survived exposure to embryonic fluid became sensitive to it if their intracellular glutathione levels were lowered. It was concluded that blatocyst contains amounts of hydrogen peroxide toxic to malignant pretrophectodermal cells and that glutathione dependent mechanisms protect inner cell mass cells with embryonic potential.

Little and Flores (1993) also concluded that hydrogen peroxide was required for apoptosis to occur. They studied a classic and dramatic example of programmed cell death: the regression of bullfrog tadpole tails during metamorphosis. Histologically and ultrastructurally, the changes in dying cells are those of apoptosis. They found that CAT, but not SOD, blocked apoptosis. They hypothesized that the source of hydrogen peroxide is cytotoxic macrophages activated during the regression process.

A key player in the apoptotic process is the protein Bcl-2. *bcl-2* is an oncogene known to cause B-cell lymphoma; *bcl-2* transgenic mice progress to high grade malignant lymphoma (McDonnell and Korsmeyer, 1991). The t(14;18) chromosomal translocation of human follicular B-cell lymphoma juxtaposes the *bcl-2* gene with the immunoglobulin heavy chain locus. The *bcl-2* immunoglobulin fusion gene is markedly deregulated resulting in inappropriately elevated levels of *bcl-2* RNA and protein. In studying the mechanism of this oncogene, it was discovered that Bcl-2 blocked the apoptotic death of a pro-B-lymphocyte cell line but had no effect on cell cycling (Hockenbery et al., 1990). Moreover, the majority of Bcl-2 protein was found on the inner mitochondrial membrane. Thus, Bcl-2 is unique among oncogenes, being localized to mitochondria and interfering with apoptosis rather than cell proliferation. Bcl-2 has been shown to inhibit apoptosis induced by a large number of stimuli including radiation, hyperthermia, growth factor withdrawal, glucocorticoids, and multiple classes of chemotherapeutic agents (Hockenbery et al., 1993).

Recently, it has been proposed that Bcl-2 acts as an antioxidant to prevent apoptosis (Hockenbery et al., 1993). Bcl-2 protected cells from hydrogen peroxide- and menadione-induced oxidative death. N-acetylcysteine, an antioxidant and inducer of GSH synthesis, inhibited apoptosis. Overexpression of GPX, but not MnSOD, inhibited apoptosis; these results suggested that hydrogen peroxide, but not superoxide, was the apoptotic agent. These results suggested that Bcl-2 scavenged some ROS. However, Bcl-2 did not affect cyanide-resistant respiration, a measure of superoxide generation, or generation of endogenous peroxides. Following an apoptotic signal, cells experienced progressive lipid peroxidation; overexpression of Bcl-2 suppressed lipid peroxidation completely. Thus, their results seem to indicate that hydrogen peroxide is the apoptotic stimulus. Bcl-2 inhibited lipid peroxidation, but not by reducing superoxide or peroxide levels. Thus, Bcl-2 appears to be a membrane protein that somehow interferes with lipid peroxidation.

In contrast to this work, Kane et al. (1993), found the Bcl-2 inhibition of neural death was associated with decreased generation of ROS. This group studied the hypothalamic neural cell line GT1-7. It was found that exposure to buthionine sulfoximine (BSO), an inhibitor of glutathione synthesis, diethyl maleate, which binds the free sulfhydryl groups of GSH, or ethacrynic acid, which depletes both cytosolic and mitochondrial GSH, all led to cell death. Cells overexpressing Bcl-2 were protected against the toxicity of these three agents and retained much higher GSH levels. It was of interest that the authors classified the cell death induced by GSH depletion as necrotic rather than apoptotic. They argue that this indicated that Bcl-2 does not inhibit apoptosis per se, but inhibits a cellular process that may result in apoptosis or necrosis. Because the elevated GSH associated with Bcl-2 might have resulted from decreased utilization of GSH, dichlorofluorescein diacetate (DCF) was used to estimate the generation of ROS in diethyl maleate- and ethacrynic acid-treated cells. DCF was thought to measure mainly hydrogen peroxide and hydroxyl radical. Treated cells demonstrated a marked decrease in fluorescence, while Bcl-2 overexpressing cells showed only a modest decline in fluorescence. The addition of the iron chelator desferrioxamine reduced ROS production and inhibited cell death. In contrast, when aminotriazole was used to inhibit CAT, GSH-depleted cells produced more ROS and cells died faster. Nitric oxide was found to play no role in death induced by GSH depletion. Bcl-2 expression enhanced the growth of mutants lacking MnSOD grown in 21% oxygen and mutants lacking CuZnSOD grown in fermentative conditions. This suggested that perhaps Bcl-2 could also protect against superoxide radical.

Sandstrom and Buttke (1993) have demonstrated that the growth of the CCRF-CEM (CEM) human CD4 T cell line in serum-free medium was cell density dependent. At cell densities less than 10^5 cells/ml, CEM cells rapidly underwent apoptosis, but at higher densities the cells remained viable due to their ability to condition the culture medium. The growth-sustaining molecule from the CEM-conditioned medium was found to be CAT. Because the level of intracellular CAT greatly exceeded that released extracellularly, it was suggested that low-density CEM cells undergo apoptosis as a result of hydrogen peroxide-mediated plasma membrane damage. This group went on to show that HIV gene expression enhanced T cell susceptibility to hydrogen peroxide- induced apoptosis (Sandstrom et al., 1993). On inoculation into serum-free medium, 8E5, a constitutive HIV-expressing T cell line, underwent apoptosis, whereas cell death was not observed with the uninfected A3.01 or latently HIV-infected 8E5L T cell lines. 8E5 survival was markedly prolonged by CAT, vitamin E, or 2-mercaptoethanol, but supplementation with ascorbic acid, glutathione, or N-acetylcysteine had no effect. 8E5 cells demonstrated lower levels of CAT activity and were more susceptible to hydrogen peroxide killing than the control cells. These authors further studied the effect of lipid hydroperoxides (Sandstrom et al., 1994). They found that HIV gene expression rendered 8E5 cells 10-fold more sensitive than uninfected parental or latently-infected cells to killing by 15-hydroperoxyeicosatetraenoic acid (15-HPETE), as well as several other hydroperoxy fatty acids. Whereas the viability of the control cells was relatively unaffected by exposure to 10 μM 15-HPETE, similarly treated 8E5 cells underwent apoptosis, as demonstrated by morphological changes and the presence of fragmented DNA. The susceptibility of 8E5 cells was attributable to their inabilty to convert 15-HPETE to 15-hydroxyeicosatraenoic acid (15-HETE) owing to a marked reduction in GPX activity in these cells.

VII. TUMOR SUPPRESSOR GENES, DOMINANT ONCOGENES, AND ANTIOXIDANT ENZYMES

Tumor suppressor genes are genes whose expression leads to suppression of the malignant phenotype. The criteria for classification as a tumor suppressor gene are two-fold: (1) loss of heterozygosity (LOH) for the gene must be demonstrated during malignant transformation, and (2) transfection and subsequent expression of the gene leads to inhibition of tumor growth. Both of these criteria have now been satisfied for MnSOD in human melanoma. LOH for MnSOD was found in approximately 33 percent of the human melanomas examined (Millikin et al., 1991). Morever, as discussed earlier, expression of MnSOD in cultured melanoma cells led to suppression of the malignant phenotype both *in vitro* and *in vivo* (Church et al., 1993). MnSOD has also been suggested to be a tumor suppressor gene in transformed human fibroblasts (Bravard et al., 1992). This designation is still not demonstrated in these cells, because although deletion of the *MnSOD* gene has been observed, the crucial transfection experiments have not been performed.

Thus, *MnSOD* is a tumor suppressor gene in certain tumor cells. Whether it is a tumor suppressor in a wide variety of tumors remains to be determined. Another question to be addressed is whether the other antioxidant enzymes are also tumor suppressor genes.

Evidence is also emerging for a relationship between the dominant oncogenes and ROS. Dominant oncogenes are genes that when expressed cause malignant transformation. These genes are found by transfecting DNA from tumor cells into normal cells and then looking for transformation. Oncogenes are thought to be derived from a normal cellular equivalent—proto-oncogenes—by processes like mutation. There are in general two classes of these dominant oncogenes: (1) cytoplasmic and membrane, and (2) nuclear. The first class of dominant oncogenes are involved in cell proliferation via signal transduction pathways. The function of the second class is still unresolved, but they appear to be involved in control of cellular functions such as differentiation by serving as regulatory genes. They are often transcription factors. Recently it has been determined that: (1) ROS induce some of these nuclear oncogenes, and (2) the function of these oncogenes is modulated by ROS.

One of the first reports of induction of proto-oncogenes by ROS was that of Shibanuma et al.(1988). They treated Balb/3T3 cells in the quiescent state with a superoxide generating system, xanthine/xanthine oxidase (X/XOD). DNA synthesis was induced 12 to 24 hours after the treatment and this response was inhibited by allopurinol, a specific inhibitor of XOD. mRNA for the proto-oncogenes *c-myc* and *c-fos* was induced 60 minutes after the addition of X/XOD. These results suggested that superoxide might be a stimulus for mitosis. Crawford and colleagues reported similarly the induction of *c-myc* and *c-fos* mRNA after X/XOD in mouse epidermal cells (1988). Nose et al. have demonstrated that exposure of mouse osteoblastic MC3T3 cells to hydrogen peroxide resulted in induction of *c-fos* and *c-jun* mRNA (1991). Exposure of primary cultures of rat proximal tubular epithelium to X/XOD provoked sequential expression of *c-fos*, *c-jun*, and *c-myc*, in this order (Maki et al., 1992). The expression of c-fos was inhibited by SOD added to the growth medium, but CAT had no effect. In this system, hydrogen peroxide alone did not induce *c-fos*.

The first report of an effect of ROS on oncogene function was that of Abate et al. (1990). They reported a redox regulation of Fos and Jun DNA-binding activity *in vitro*. The proto-oncogenes *fos* and *jun* function cooperatively as inducible transcription factors in signal transduction processes. Their protein products, Fos and Jun, form a heterodimeric complex that interacts with the DNA regulatory element activator protein-1 (AP-1) binding site. DNA binding of the Fos-Jun heterodimer was modulated by reduction-oxidation (redox) via sulfhydryl reagents of a single conserved cysteine residue in the DNA-binding domain of the two proteins. This study also identified a nuclear protein that reduced Fos and Jun and stimulated DNA-binding *in vitro*. Thus, reduction of the conserved cysteine residue by chemical reducing agents or by a nuclear redox-factor (later termed Ref-1) stimulated DNA binding *in vitro*, whereas oxidation or chemical modification of the cysteine residue had an inhibitory effect on DNA binding activity. Xanthoudakis et al. (1992) have shown that the protein product of the *ref-1* gene stimulates the DNA binding activity of Fos-Jun heterodimers, Jun-Jun homodimers, and HeLa cell AP-1 proteins as well as several other transcription factors including NF-KappaB, Myb, and members of the ATF/CREB family. Some of these transcription factors are proto-oncogenes, such as *myb*, while others have not been demonstrated yet to be oncogenes. Immunodepletion analysis indicated that Ref-1 was the major AP-1 redox activity in HeLA nuclear extracts. Ref-1 is a bifunctional protein; it also possesses an apurinic/apyrimidinic (AP) endonuclease repair activity. The redox and DNA repair activities of Ref-1 can in part be distinguished.

The importance of this redox regulation on transformation by the *fos* oncogene has recently been demonstrated. To determine whether redox control affects the function of Fos, Okuno et al. (1993) compared the properties of retroviral vectors expressing either a truncated Fos protein (F118-211) or a truncated Fos protein in which the critical cysteine was replaced by a serine (FC154S). In infected chicken embryo fibroblasts (CEFS), both vectors expressed similar levels of Fos, which formed heterodimers with Jun at equivalent efficiencies. Extracts from cells expressing FC154S showed a threefold increase in AP-1 DNA- binding activity compared to cells expressing F118-211. This enhanced binding activity was resistant to treatment with the oxidizing agent diamide. Infection of CEFs by virus expressing FC154S resulted in increased numbers of transformed colonies and an increase in colony size compared with those obtained following infection by virus expressing F118-211. These results suggest that redox regulation may limit the total level of functional Fos–Jun complexes and, more importantly, that escape from redox control enhances the transforming activity of *fos*. The authors noted that only a three-fold increase in the DNA-binding activity of Fos–Jun complexes was sufficient to enhance transformation frequency and the transformed phenotype.

Similar observations have been made for the *v-myb* oncogene. The *c-myb* gene is the proto-oncogene of *v-myb*—the genetic component of two retroviruses, avian leukemic virus E26 and avian myeloblastosis virus AMV. The *c-myb* gene is involved in the regulation of proliferation and/or differentiation of hematopoietic progenitor cells. Down regulation or antisense inhibition of *c-myb* transcripts inhibits proliferation and leads to differentiation. Recently it has been shown that reduction of a conserved cysteine is essential for c-Myb DNA-binding (Guehmann et al., 1992). The v-Myb protein transforms myelomonocytic cells. There are seven cysteines in v-Myb. One of these cysteines, amino acid 65, was found to be very important in transformation by *v-myb*. When cysteine 65 was replaced with a serine, the mutant protein neither trans-activated transcription nor transformed myeloid cells, although it was transported into the nucleus. This cysteine seemed to be required for high-affinity DNA binding.

Two other dominant oncogenes have been recently shown to be redox regulated. It has been demonstrated that oncogenic conversion of Ets affects redox modulation (Wasylyk and Wasylyk, 1993). Even more important, the p53 protein, which is abnormal in a very large fraction of human cancers, has been reported to be redox modulated in both its conformation and DNA binding (Hainut and Milner, 1993). This report suggests that the function of wild-type p53 could be inhibited by ROS and stimulated by the counteracting cellular reducing response.

Thus, the early data suggests that certain of the dominant oncogenes are redox sensitive and mutations in the DNA that regulates this redox control leads to cellular transformation. What relationship does this have to MnSOD? The molecular structure for the rat *MnSOD* gene has been published by Ho et al. (1991). They found that the rat *MnSOD* gene promoter contained one copy of the AP-1 (c-Fos and c-Jun binding site) binding motif. Thus, we hypothesize that Fos and Jun in part regulate *MnSOD* transcription. Tumor cells are low in *MnSOD* either because the promoter is altered so that transcription factors like Fos-Jun do not bind efficiently or the transcription factors themselves are altered so that they do not bind properly. Thus, in this specific example, *MnSOD* could be low because of a mutation in its promoter or because the transcription factors Fos-Jun are deregulated so they no longer bind to the promoter. Fos-Jun could be deregulated because of mutations in either the redox or phosphorylation domains, as both domains have been shown to cause loss of DNA binding capacity (Oehler et al., 1993).

As discussed earlier, *MnSOD* has been demonstrated to be a tumor suppressor gene in human melanoma. Thus, we extrapolate these findings to postulate a general principle: certain of the nuclear proto-oncogenes regulate the transcription of certain of the tumor suppressor genes. Malignancy can be caused by loss of the tumor suppressor gene function, either by loss of the dominant tumor suppressor allele or by lack of regulation of the suppressor gene by the dominant oncogene. Lack of regulation occurs because of improper binding of the dominant oncogene to the promoter of the tumor suppressor gene or improper functioning once bound. This theory then for the first time ties the function of dominant oncogenes to tumor suppressor genes.

There is also probably feedback from the tumor suppressor to the dominant oncogene. Since ROS induce *fos* and *jun*, elevation of SOD could lead to repression of *fos* and *jun* due to reduction in levels of ROS. Thus, in general, certain of the dominant oncogenes may modulate certain tumor suppressor genes, which may in turn modulate the dominant oncogenes.

VIII. Tat PROTEIN OF HIV TYPE 1 REPRESSES MnSOD

It has been demonstrated that human immunodeficiency virus (HIV)-infected HUT-78 cells have decreased levels of MnSOD and lose their ability to induce this enzyme after exposure to tumor necrosis factor (Wong et al., 1991). Flores et al. (1993), as discussed in this volume, have studied this inhibition further. They have observed the effect of one of the proteins, Tat, produced by HIV in cells (Flores et al., 1993). Several of these HIV gene products are important in the regulation of viral expression. The Tat (tran-activating transcription activator) and Rev proteins are essential for virus replication. Tat binds to a sequence, termed the transactivation response element (TAR), located downstream from the site of transcriptional initiation in the viral long terminal repeat. Transcripts containing the TAR RNA sequence adopt a stable stem-loop structure to which Tat binds specifically and tightly. Some investigators have proposed that Tat works by modulating mRNA stability, others have suggested it increased the rate of viral transcription, and still others have proposed that Tat acts at both the transcriptional and post-transcriptional level.

Tat is important to our discussion of transformation because in addition to regulating HIV gene expression and replication, it also influences cellular phenotype by modulating the expression of certain cellular genes. In particular, Tat expression leads to transformation of primary keratinocyte cultures (Kim et al., 1992). The tumorigenic potential of Tat is further demonstrated in transgenic mice by the induction of dermal lesions resembling Kaposi's sarcoma (Vogel et al., 1988), alterations in the expression of angiogenic factors after introduction of Tat-producing cells into nude mice (Huber et al., 1992), and growth promotion of cells derived from Kaposi's sarcoma lesions of AIDS patients (Ensoli et al., 1990). Thus, Tat is a transforming protein.

Tat has also been demonstrated to inhibit the expression of MnSOD (Flores et al., 1993). This group stably transfected the *tat* gene into the HeLa tumor cell line. The parental HeLa cells expressed nearly equivalent amounts of CuZnSOD and MnSOD. Those cells expressing the Tat protein contained 52% less MnSOD activity than parental cells, while CuZnSOD activity was unchanged. The steady-state levels of MnSOD mRNA, but not CuZnSOD mRNA, were also lower in the HeLa-tat line than in the parental line. The Tat-overexpressing cells demonstrated increased oxidative stress, as evidenced by increased carbonyl proteins and decreased total cellular sulfhydryl content. Most interesting, this group demonstrated a direct interaction between Tat protein and *MnSOD* gene transcripts. It could be that this RNA binding causes the repression of MnSOD observed. It could do this by decreasing RNA stability and thus leading to less protein being translated. In this regard, it is interesting that Fazzone et al. (1993) have recently reported the existence of a MnSOD mRNA-binding protein. They find this protein is redox-sensitive and decreases during development. They hypothesize that this MnSOD mRNA-binding protein is a negative regulator of MnSOD mRNA stability. In other words, the more of this protein that binds the the MnSOD-mRNA, the less is the stability of the mRNA and the less MnSOD protein made. Tat could act in a similar fashion. The levels of this normal mRNA binding protein could explain the increases in MnSOD levels usually observed during development.

Whatever the mechanism of MnSOD inhibition caused by Tat, the inhibition suggests again that low levels of MnSOD are related to transformation. The lowering of MnSOD by Tat is especially interesting because it has been observed for years that many virally transformed cells have low levels of MnSOD. A widely studied example is SV40-transformed fibroblasts. We are currently transfecting virally-transformed cells to determine the effect on the malignant phenotype. It would also be interesting to do such transfection experiments in the *tat*-transfected cells.

IX. PROSPECTS FOR THE FUTURE

If these ideas prove correct, they have wide implications for both the prevention and treatment of cancer. With regards to prevention, this work suggests that further testing of the usefulness of natural and synthetic antioxidants to prevent cancer has merit. Perhaps workers occupationally exposed to carcinogens could benefit from antioxidant supplements in the diet. Along the same lines, the usefulness in differentiation/normalization therapy of cancer of liposomal and PEG-SODs, synthetic SODs, and molecular biological techniques to increase SOD in tumor cells should be investigated.

ACKNOWLEDGMENTS

The authors would like to thank Mrs. Susan Barnett for her help in manuscript preparation. This work was supported in part by NIH grant CA 41267.

REFERENCES

Abate, C., Patel, L., Rauscher III, F.J., & Curran, T. (1990). Redox regulation of Fos and Jun DNA-binding activity *in vitro*. Science 249, 1157–1161.

Allen, R.G., Balin, A.K., Reimer, R.J., Sohal, R.S., & Nations, C. (1988). Superoxide dismutase induces differentiation in microplasmodia of the slime mold *Physarum polycephalum*. Arch. Biochem. Biophys. 261, 205–211.

Armato, U., Andreis, P.G., & Romano, F. (1984). Exogenous Cu,Zn- superoxide dismutase suppresses the stimulation of neonatal rat hepatocytes' growth by tumor promoters. Carcinogenesis 5, 1547–1555.

Asoh, K.-I., Watanabe, Y., Mizoguchi, H., Mawatari, M., Ono, M., Kohno, K., & Kuwano, M. (1989). Induction of manganese superoxide dismutase by tumor necrosis factor in human breast cancer MCF-7 cell line and its TNF-resistant variant. Biochem. Biophys. Res. Commun. 162, 794–801.

Beckman, B.S., Balin, A.K., & Allen, R.G. (1989). Superoxide dismutase induces differentiation of erythroleukemia cells. J. Cell. Physiol. 139, 370–376.

Birnboim, H.C. (1981). Factors which affect DNA strand breakage in human leukocytes exposed to a tumor promoter, phorbol myristate acetate. Can. J. Physiol. Pharmacol. 60, 135–366.

Bize, I.B., Oberley, L.W., & Morris, H.P. (1980). Superoxide dismutase and superoxide radical in the Morris hepatomas. Cancer Res. 40, 3686–3693.

Bravard, A., Sabatier, L., Hoffschir, F., Luccioni, C., & Dutrillaux, B. (1992). SOD2: A new type of tumor suppressor gene? Int. J. Cancer 51, 475–480.

Burdon, R.H., & Rice-Evans, C. (1989). Free radicals and the regulation of mammalian cell proliferation. Free Radical Res. Commun. 6, 345.

Ceballos, I., Delabar, J.M., Nicole, A., Lynch, R.E., Hallewell, R.A., Kamoun, P., & Sinet, P.M. (1988). Expression of transfected human CuZn superoxide dismutase gene in mouse L cells and NS20Y neuroblastoma cells induces enhancement of glutathione peroxidase activity. Biochim. Biophys. Acta 949, 58–64.

Cerutti, P.A. (1985). Prooxidant states and cancer. Science 227, 375–381.

Cerutti, P.A., Emerit, I., & Amstad, P. (1983). Membrane-mediated chromosomal damage. In: Genes and Proteins in Oncogenesis (Weinstein, I.B., ed.), pp. 5–7. Academic Press, New York.

Church, S.L., Grant, J.W., Ridnour, L.A., Oberley, L.W., Swanson, P.E., Meltzer, P.S., & Trent, J.M. (1993). Increased manganese superoxide dismutase expression suppresses the malignant phenotype of human melanoma cells. Proc. Natl. Acad. Sci. USA, 90, 3113–3117.

Craven, P.A., Pfanstiel, J., & DeRubertis, F.R. (1986). Role of reactive oxygen in bile salt stimulation of colonic epithelial proliferation. J. Clin. Invest. 77, 850–859.

Crawford, D., Zbinden, I., Amstad, P., & Cerutti, P. (1988). Oxidant stress induces the proto-oncogenes *c-fos* and *c-myc* in mouse epidermal cells. Oncogene 3, 27–32.

Cunningham, M.L., & Lokesh, B.R. (1983). Superoxide anion generated by potassium superoxide is cytotoxic and mutagenic to Chinese hamster ovary cells. Mutat. Res. 121, 299–304.

Cunningham, M.L., Ringrose, P.S., & Lokesh, B.R. (1984). Inhibition of genotoxicity of bleomycin by superoxide dismutase. Mutat. Res., 135, 199–202.

Czaja, M.J., Schilsky, M.L., Xu, Y., Schmiedeberg P., Compton, A., Ridnour, L., & Oberley, L.W. (1994). Induction of MnSOD gene expression in a hepatic model of TNF-α toxicity does not result in increased protein. Am. J. Physiol., 266, G737-G744.

Elroy-Stein, O., Berstein, Y., & Groner, Y. (1986). Overproduction of human Cu/Zn-superoxide dismutase in transfected cells: Extenuation of paraquat-mediated cytotoxicity and enhancement of lipid peroxidation. EMBO J. 5, 615–622.

Ensoli, B., Barillari, G., Salahuddin, S.Z., Gallo, R.C., & Wong-Staal, F. (1990). Tat protein of HIV-1 stimulates growth of cells derived from Kaposi's sarcoma lesions of AIDS patients. Nature 345, 84–86.

Fazzone, H., Wangner, A., & Clerch, L.B. (1993). Rat lung contains a developmentally regulated manganese superoxide dismutase mRNA-binding protein. J. Clin. Invest. 92, 1278–1281.

Fernandez-Pol, J.A., Hamilton, P.D., & Klos, D.J. (1982). Correlation between the loss of the transformed phenotype and an increase in superoxide dismutase activity in a revertant subclone of sarcoma virus-infected mammalian cells. Cancer Res. 42, 609–617.

Flores S.C., Marecki, J.C., Harper, K.P., Bose, S.K., Nelson, S.K., & McCord, J.M. (1993). Tat protein of human immunodeficiency virus type 1 represses expression of manganese superoxide dismutase in HeLa cells. Proc. Natl. Acad. Sci. USA 90, 7632–7636.

Fujii, J., & Taniguchi, N. (1991). Phorbol ester induces manganese-superoxide dismutase in tumor necrosis factor-resistant cells. J. Biol. Chem.266, 23142–23146.

Galeotti, T. Wohlrab, H., Borrello, S., & Emilia de Leo, M. (1989). Messenger RNA for manganese and copper-zinic superoxide dismutases in hepatomas: Correlation with degree of differentiation. Biochem. Biophys. Res. Commun. 165, 581–589.

Goldstein, B.D., Witz, G., Amoruso, M., Stone, D.S., & Troll, W. (1981). Stimulation of human polymorphonuclear leukocyte superoxide anion radical production by tumor promoters. Cancer Lett. 11, 257–262.

Gramzinski, R.A., Parchment, R.E., & Pierce, G.B. (1990). Evidence linking programmed cell death in the blastocyst to polyamine oxidation. Differentiation 43, 59–65.

Guehmann, S., Vorbrueggen, G., Kalkbrenner, F., & Moelling, K. (1992). Reduction of a conserved Cys is essential for Myb DNA-binding. Nucleic Acids Res. 20, 2279–2286.

Hainaut, P., & Milner, J. (1993). Redox modulation of p53 conformation and sequence-specific DNA binding *in vitro*. Cancer Res. 53, 4469–4473.

Himeno, T., Watanabe, N., Yamuchi, N., Maeda, M., Okamoto, T., Tsuiji, N., Tsusi, Y., Akiyama, S., Sasaki, H., & Niitsu, Y. (1992). Induction of synthesis of manganous superoxide dismutase in L-M (pNTnF) cells carrying an inducible TNF gene. Int. J. Cancer 50, 458–462.

Ho, Y.-S., Howard, A.J., & Crapo, J.D. (1991). Molecular structure of a functional rat gene for manganese-containing superoxide dismutase. Am. J. Resp. Cell. Mol. Biol. 4, 278–286.

Hockenbery, D., Nunez, G., Milliman C., Schreiber, R.D., & Korsmeyer, S.J. (1990). Bcl-2 is an inner mitochondrial membrane protein that blocks programmed cell death. Nature 348, 334–336.

Hockenbery, D.M., Oltval, Z.N., Yin, X.-M., Milliman, C.L., & Korsmeyer, S.J. (1993). Bcl-2 functions in an antioxidant pathway to prevent apoptosis. Cell 75, 241–251.

Huber, B.E., Richards, C.A., Martin, J.L., & Wirth, P.J. (1992). Alterations in human angiogenesis associated with stable expression of the HIV tat gene. Mol. Carcinogen. 5, 293–300.

Kane, D.J., Sarafian, T.A., Anton, R., Hahn, H., Gralla, E.B., Valentine, J.S., Ord, T., & Bredesen, D.E. (1993). Bcl-2 inhibition of neural death: decreased generation of reactive oxygen species. Science 262, 1274–1277.

Kawaguchi, T., Takeyase, A., Matsunobu, K., Uda, T., Ishiiizawa, M., Suzuki, K., Nishiura, T., Isikawa, M., & Taniguchi, N. (1990). Stimulation of Mn-superoxide dismutase expression by tumor necrosis factor-α: quantitative determination of Mn-SOD protein levels in TNF-α resistant and sensitive cells by ELISA. Biochem. Biophys. Res. Commun. 171, 1378–1386.

Kelner, M., & Bagnell, R. (1990). Alteration of endogenous glutathione peroxidase, manganese superoxide dismutase, and glutathione transferase activity in cells transfected with a copper-zinc superoxide dismutase expression vector. J. Biol. Chem. 265, 10872–10875.

Kensler, T.W., & Trush, M.A. (1984). Role of oxygen radicals in tumor promotion. Environ. Mutagen. 6, 593–616.

Kim, C.M., Vogel, J., Jay, G., & Rhim, J.S. (1992). The HIV *tat* gene transforms human keratinocytes. Oncogene 7, 1525–1529.

Konstantinov, A.A., Peskin, A.V., Popova, E.Y., Knomutov, G.B., & Runge, E.K. (1987). Superoxide generation by the respiratory chain of tumor mitochondria. Biochim. Biophys. Acta 894, 1–10.

Little, G.H., & Flores, A. (1993). Inhibition of programmed cell death by catalase and phenylalanine methyl ester. Comp. Biochem. Physiol. 105A, 79–83.

Loven, D.P., Guernsey, D.L., & Oberley, L.W. (1984). Transformation affects superoxide dismutase activity. Int. J. Cancer 33, 783–786.

Maki, A., Berezesky, I.K., Fargnoli, J., Holbrook, N.J., & Trump, B.F. (1992). Role of [Ca^{2+}] in induction of *c-fos*, *c-jun*, and *c-myc* mRNA in rat PTE after oxidative stress. FASEB J. 6, 919–924.

Marklund, S. (1992). Regulation by cytokines of extracellular superoxide dismutase and other superoxide dismutase isoenzymes in fibroblasts. J. Biol. Chem. 267, 6696–6701.

Marlens, F., Nicole, A., & Sinet, P.M. (1985). Lowered level of translatable messenger RNAs for manganese superoxide dismutase in human fibroblasts transformed by SV40. Biochem. Biophys. Res. Commun. 129, 300–305.

Masuda, A., Longo, D.L., Kobayashi, Y., Appella, E., Oppenheim, J.J., & Matsushima, K. (1988). Induction of mitochondrial superoxide dismutase by interleukin-1. FASEB J. 2, 3087–3091.

McCord, J.M., Keele, Jr., B.B., & Fridovich, I. (1971). An enzyme based theory of obligate anaerobiosis: the physiological function of superoxide dismutase. Proc. Natl. Acad. Sci. USA 68, 1024–1027.

McCormick, M.L., Oberley, T.D., Elwell, J.H., Oberley, L.W., Sun, Y., & Li, J.J. (1991). Superoxide dismutase and catalase levels during estrogen-induced renal tumorigenesis in renal tumors and their autonomous variants in the Syrian hamster. Carcinogenesis, 12, 977–983.

McDonnell, T.J., & Korsmeyer, S.J. (1991). Progression from lymphoid hyperplasia to high grade malignant lymphoma in mice transgenic for the t(14;18). Nature 349, 254–256.

Miller, R.C., Osmak, R., Zimmerman, M., & Hall, E.J. (1982) Sensitizers, protectors, and oncogenic transformation *in vitro*. Int. J. Radiat. Oncol. 8, 771–775.

Millikin, D., Meese, E., Vogelstein, B., Witkowski, C., & Trent, J. (1991). Loss of heterozygosity for loci on the long arm of chromosome 6 in human malignant melanoma. Cancer Res. 51, 5449–5453.

Murrell, G.A.C., Francis, M.J.O., & Bromley, L. (1990). Modulation of fibroblast proliferation by oxygen free radicals. Biochem. J. 265, 659–665.

Nose, K., Shibanuma, M., Kikuchi, K., Kageyama, H., Sakiyama, S., & Kuroki, T. (1991). Transcriptional activation of early-response genes by hydrogen peroxide in a mouse osteoblastic cell line. Eur. J. Biochem. 201, 99–106.

Oberley, L.W. (1982). Superoxide dismutase and cancer. In: Superoxide Dismutase, Vol. II, (Oberley, L.W., Ed.), Ch. 6, CRC Press, Boca Raton, FL.

Oberley, L.W. , Bize, I.B., Sahu, S.K., Leuthauser, S.W.H., & Gruber, H.E. (1978). Superoxide dismutase activity of normal murine liver, regenerating liver, and H6 hepatoma. J. Natl. Cancer Inst. 61, 375–379.

Oberley, L.W., & Buettner, G.R. (1979). Role of superoxide dismutase in cancer: A review. Cancer Res. 39, 1141–1149.

Oberley, L.W., Kasemset St.Clair, D., Autor, A.P., & Oberley, T.D. (1987). Increase in manganese superoxide dismutase activity in the mouse heart after X-irradiation. Archives Biochem. Biophys. 254, 69–80.

Oberley, L.W., McCormick, M.L., Sierra, E., & Kasemset St.Clair, D. (1989a). Manganese superoxide dismutase in normal and transformed human embryonic lung fibroblasts. Free Radical Biol. Med. 6, 379–384.

Oberley, L.W., & Oberley, T.D. (1984). The role of superoxide dismutase and gene amplification in carcinogenesis. J. Theor. Biol., 106, 403–422.

Oberley, L.W., & Oberley, T.D. (1986). Free Radicals, cancer, and aging. In: Free Radicals, Aging, and Degenerative Diseases (Johnson, Jr., J.E., Walford, R., Harmon, D., & Miquel, J., Eds.), pp. 325–371. Alan R. Liss, Inc.

Oberley, L.W., Ridnour, L.A., Sierra-Rivera, E., Oberley, T.D., & Guernsey, D.L. (1989b). Superoxide dismutase activities of differentiating clones from an immortal cell line. J. Cell. Physiol. 138, 50–60.

Oberley, T. (1985). The possible role of reactive oxygen metabolites in cell division. In: Superoxide Dismutase, Vol. III (Oberley, L.W., Ed.), pp. 83–97. CRC Press, Boca Raton, FL.

Oberley, T.D., Allen, R.G., Schultz, J.L., & Lauchner, L.J. (1991). Antioxidant enzymes and steroid-induced proliferation of kidney tubular cells. Free Radical Biol. Med. 10, 79-83.

Oberley, T.D., Oberley, L.W., Slattery, A.F., Lauchner, L.J., & Elwell, J.H. (1990). Immunohistochemical localization of antioxidant enzymes in adult Syrian hamster tissues and during kidney development. Am. J. Pathol. 137, 199–214.

Oberley, T.D., Sempf, J.M., Oberley, M.J., McCormick, M.L., Muse, K.E., & Oberley, L.W. (1994). Immunogold analysis of antioxidant enzymes in human renal cell carcinoma. Virchows Archives, 424, 155-164.

Oehler, T., Pintzas, A., Stumm, S., Darling, A., Gillespie, D., & Angel, P. (1993). Mutation of a phosphorylation site in the DNA binding domain is required for redox-independent transactivation of AP1-dependent genes by v-Jun. Oncogene 8, 1141–1147.

Okuno, H., Akahori, A., Sato, H., Xanthoudakis, S., Curran, T., & Iba, H. (1993). Escape from redox regulation enhances the transforming activity of Fos. Oncogene 8, 695–701.

Ono, M., Kohda, H., Kawaguchi, T., Ohhira, M., Sekiya, C., Namiki, M., Takeyasu, A., & Taniguchi, N. (1992). Induction of Mn-superoxide dismutase by tumor necrosis factor, interleukin-1 and interleukin-6 in human hepatoma cells. Biochem. Biophys. Res. Commun, 182, 1100–1107.

Pang, X.-P., Ross, R.S., Park, M., Juillard, G.J.F., Stanley, T.M., & Hershman, J.M. (1992). Tumor necrosis factor-α activates nuclear factor KB and induces manganous superoxide dismutase and phosphodiesterase mRNA in human papillary thyroid carcinoma cells. J. Biol. Chem. 267, 12826–12830.

Parchment, R.E., Gramzinski, R.A., & Pierce, G.B. (1990). Neoplastic embryoid bodies of embryonal carcinoma C44 as a source of blastocele-like fluid. Differentiation 43, 51–58.

Petkau, A., Monasterski, L.G., Kelly, K., & Friesen, H.G. (1977). Modification of superoxide dismutase in rat mammary carcinoma. Res. Commun. Chem. Pathol., & Pharmacol. 17, 125–132.

Pierce, G.B., Parchment, R.E., & Lewellyn, A.L. (1991). Hydrogen peroxide as a mediator of programmed cell death in the blatocyst. Differentiation 46, 181–186.

Sandstrom, P.A., & Buttke, T.M. (1993). Autocrine production of extracellular catalase prevents apoptosis of the human CEM T cell line in serum-free media. Proc. Natl. Acad. Sci. USA 90, 4708- 4712.

Sandstrom, P.A., Roberts, B., Folks, T.M., & Buttke, T.M. (1993). HIV gene expression enhances T cell susceptibility to hydrogen peroxide-induced apoptosis. AIDS Res. Human Retroviruses 9, 1107–1113.

Sandstrom, P.A., Tebbey, P.W., Van Cleave, S., & Buttke, T.M. (1994). Lipid hydroperoxides induce apoptosis in T cells displaying a HIV-associated glutathione peroxidase deficiency. J. Biol. Chem. 269, 798–801.

Shibanuma, M., Kuroki, T., & Nose, K. (1988). Induction of DNA replication and expression of proto-oncogene c-myc and c-fos in quiescent Balb/3T3 cells by xanthine/xanthine oxidase. Oncogene 3, 17–21.

Springer, E.L. (1980). Comparative study of the cytoplasmic organelleles of epithelial cell lines derived from human carcinomas and nonmalignant tissues. Cancer Res. 40, 803–817.

St.Clair, D.K., & Holland, J.C. (1991). Complementary DNA encoding human colon cancer manganese superoxide dismutase and the expression of its gene in human cells. Cancer Res. 51, 939–943.

St.Clair, D.K., & Oberley, L.W. (1991). Manganese superoxide dismutase expression in human cancer cells: a possible role of mRNA processing. Free Radical Res. Commun. 12-13, 771–778.

St.Clair, D.K., Oberley, T.D., Muse, K.E., & St.Clair, W.H. (1994). Expression of manganese superoxide dismutase promotes cellular differentiation. Free Radical Biol. Med., in press.

St.Clair, D.K., Wan, X.S., Oberley, T.D., Muse, K.E., & St.Clair, W.H. (1992). Suppression of radiation-induced neoplastic transformation by overexpression of mitochondrial superoxide dismutase. Molec. Carcinogenesis 6, 238–242.

Sun, Y., Colburn, N.H., & Oberley L.W. (1993b). Decreased expression of manganese superoxide dismutase mRNA and protein after immortalization and transformation of mouse liver cells. Oncology Res. 5, 127–132.

Sun, Y., Li, Y., & Oberley, L.W. (1988). Superoxide dismutase activity during dimethylhydrazine colon carcinogenesis and the effects of cholic acid and indole. Free Radical Res. Commun. 4, 299–310.

Sun, Y., Oberley, L.W., Oberley, T.D., Elwell, J.H., & Sierra-Rivera, E. (1993a). Lowered antioxidant enzymes in spontaneously transformed embryonic mouse liver cells in culture. Carcinogenesis 14, 1457–1463.

Vogel, J., Hinrichs, S.H., Reynolds, R.K., Luciw, P.A., & Jay, G. (1988). The HIV tat gene induces dermal lesions resembling Kaposi's sarcoma in transgenic mice. Nature 335, 606–611.

Warner, B .B., Burhans, M.S., Clark, J.C., & Wispe, J.R. (1991). Tumor necrosis factor-α increases Mn-SOD expression: protection against oxidant injury. Am. J. Physiol. 260, L296–L301.

Wasylyk, C., & Wasylyk, B. (1993). Oncogenic conversion of Ets affects redox regulation *in vivo*. Nucleic Acids Res. 21, 523–529.

Williams, G.T. (1991). Programmed cell death: Apoptosis and oncogenesis. Cell 65, 1097–1098.

Wong, G.H.W., Elwell, J.H., Oberley, L.W., & Goeddel, D.V. (1989). Manganous superoxide dismutase is essential for cellular resistance to cytotoxicity of tumor necrosis factor. Cell 58, 923–931.

Wong, G.H.W., & Goeddel, D.V. (1988). Induction of manganous superoxide dismutase by tumor necrosis factor: Possible protective mechanism. Science 242, 941–944.

Wong, G.H.W., McHugh, T.M., Weber, R., & Goeddel, D.V. (1991). Tumor necrosis factor alpha selectively sensitizes human immunodeficiency virus-infected cells to heat and radiation. Proc. Natl. Acad. Sci., USA 88, 4372–4376.

Xanthoudakis, S., Miao, G., Wang, F., Pan, Y.-C. E., & Curran, T. (1992). Redox activation of Fos-Jun DNA binding activity is mediated by a DNA repair enzyme. EMBO J, 11, 3323–3335.

Zimmerman, R., & Cerutti, P. (1984) Active oxygen acts as a promoter of transformation in mouse embryo C3H/10T-1/2/C18 fibroblasts. Proc. Natl. Acad. Sci. USA 81, 2085–2087.

OXIDATIVE STRESS AND HUMAN IMMUNODEFICIENCY VIRUS

Sonia C. Flores and Joe M. McCord

I. OXIDATIVE BALANCE AND DISEASE

Throughout the course of this book, the reader has been exposed to a variety of pathologies whose etiology or progression is free-radical mediated. Rather than

Advances in Molecular and Cellular Biology
Volume 25, pages 71-94.

ISBN: 0-7623-0379-4

causing diseases directly, free radicals in most cases lead to an alteration of redox homeostasis which increases cellular sensitivity to additional insults. What results is a "dose-response" to stress. Some of the active oxygen species which participate in alterations of this balance are the superoxide radical ($O_2^{\bullet -}$), hydrogen peroxide (H_2O_2), the hydroxyl radical (HO·), and hypochlorous acid (HOCl). These are produced by activated phagocytes (Babior, 1978) during infections or other inflammatory conditions, or by ischemia/reperfusion injury (McCord, 1985). Of these species, the HO· is the most reactive. In spite of its low reactivity, the superoxide radical is responsible for the rapid inactivation of enzymes which contain iron–sulfur centers. It is the release of iron from these centers or from ferritin and its participation in iron-catalyzed lipid peroxidation which accounts for much of the damaging effects of the superoxide radical. Cells which accumulate lipid hydroperoxides exhibit altered membranes and impaired functional ability. Fortunately, there are cellular mechanisms which have evolved to prevent or repair this damage. Catalase eliminates H_2O_2, glutathione peroxidase eliminates fatty acid hydroperoxides, and the superoxide dismutases eliminate $O_2^{\cdot -}$. Other cellular antioxidants whose decreased levels indicate increased oxidative stress are glutathione (GSH), ascorbic acid (vitamin C), vitamin E, β-carotene, glucose-6-phosphate dehydrogenase, and glutathione peroxidase. Increases in prooxidants or decreases in antioxidants alter this balance with similar cellular outcomes.

II. VIRUS INFECTIONS AND OXIDATIVE STRESS

As soon as a virus infects a host, cellular metabolism is redirected in order to provide the substrates, energy, and machinery for replication of the virus genome. Thus, a competition for metabolic substrates results. Interactions between the virus and the host as well as the host's response to the infection will determine the outcome of the disease. In general, virus replication initiates changes in cells that can lead to cytolysis or alterations in the appearance and functional properties of the cell. Viruses take advantage of cellular transcriptional or replicative changes by using cellular DNA and RNA polymerases and nucleotide precursors. If a cell is quiescent, then this cell is an unsuitable host. The virus has two choices: either it infects cells which are in a mitotic state, or it infects a quiescent cell and then stimulates it to divide, resulting in a transformed cell. Oncogenic retroviruses cause cancer by capturing cellular genes that play an essential role during cell proliferation and/or development and removing these from their appropriate controls. When these retroviruses infect other cells, the unregulated gene is then expressed inappropriately.

An alternative mechanism for removing a mitotic block is the induction of a prooxidant state which is interpreted by mitotically competent cells as a signal to divide. This indirect response to virus infection is probably an evolutionary adaptation to inflammation, one goal of which is to initiate a process of wound healing and therefore localized cellular proliferation. Thus, viruses, like their host cells themselves, are able to replicate very rapidly when the redox status of the cell is shifted towards the oxidizing side. Some examples of this phenomenon include the increased production of free radicals by mouse lung epithelial cells upon influenza virus infection (Maeda and Akaike, 1991), partly due to increased activity of the superoxide-generating enzyme xanthine oxidase (XO) (Akaike et al., 1990) and partly due to depressed antioxidant levels (Hennet et al., 1992). Treatment with allopurinol (an inhibitor of XO) in conjunction with SOD improves the survival of these mice (Akaike et al., 1990; Maeda and Akaike, 1991). Neutrophils, which are one of the first lines of defense against bacterial infections, undergo an oxidative burst when stimulated appropriately. This oxidative burst may be triggered directly by the infectious agent itself, or indirectly, by some secondary factor produced during the infection. Incubation of isolated neutrophils in the presence of influenza virus enhances their superoxide production after an activating stimulus is applied (Arora and Henrichon, 1994; Maeda and Akaike, 1991). Increased xanthine oxidase after cytomegalovirus infection correlates with the severity of the pathological changes observed in the lung (Ikeda et al., 1992), and in developing brain and spinal cord (Boyes et al., 1989). In these cases, the damage is not caused by the virus but by the invading inflammatory cells.

Interferon-inducing agents such as RNA viruses also induce xanthine oxidase, an effect which is inhibited by the reducing agent dithiothreitol (DTT), suggesting that the conversion of the nonsuperoxide producing dehydrogenase form to the superoxide-producing oxidase form is inhibited (Deloria et al., 1985). Sendai viruses stimulate the generation of chemiluminescence within a few seconds after addition to a suspension of mouse spleen cells. Because of their membrane-altering properties, these viruses probably stimulate the signal transduction pathways on the target cell that result in superoxide production (Peterhans et al., 1988). Perturbations of cellular membranes by either chemical or infectious agents may trigger a cascade of events that ends in either proliferation or apoptosis. The activation of intracellular superoxide-generating systems may be an intermediate step in either one of these pathways.

SV40 transformed cells have considerably less MnSOD activity than the normal parental cells, regardless of growth conditions (Yamanaka and Deamer, 1974). This decrease correlates with loss of chromosome 6q, which contains the MnSOD gene (Bravard et al., 1992). Because of their ability to suppress the malignant phenotype, the SODs have been called tumor suppressor genes (Bravard et al., 1992). Thus, by inhibiting the expression of antioxidant enzymes directly, viral antigens may promote transformation.

In many respects, initial herpes simplex virus infections are like any other virus infection: exposure to the virus leads to an acute viremia which is then subsequently cleared by the immune system. Nevertheless, herpes viruses have the ability to become latent, usually in sensory ganglia of the nervous system. This period of latency varies from individual to individual. Eventually, some extracellular stimulus which may be sunlight exposure, stress, fever or infection, causes reactivation of the virus. All of these triggers result in an increased production of active oxygen species, suggesting a correlation between free radical production and herpesvirus reactivation. Even though the nature of the intracellular signals that transduce the information from the initial stimulus to the final reactivation are unknown, a specific cellular factor has been postulated (Ralph et al., 1994). This cellular factor appears to be up-regulated as a result of ultraviolet irradiation.

III. HUMAN IMMUNODEFICIENCY VIRUS AND OXIDATIVE STRESS

The etiology of AIDS, with a few exceptions, is not questioned. What is yet not clear is how does a patient progress from HIV-positivity to full-blown AIDS. Furthermore, how does one explain the paradox that there is a progressive loss of T-lymphocytes even though only a small percentage are actually infected? Even though it is assumed that AIDS is an immunological disease resulting from destruction of the T-cells and the consequent failure of the immune system, it is clear that after more than ten years of the epidemic, there is still no cure in sight. Therefore, alternative explanations and different avenues of research have to be devised. Some scientists now believe that some other co-factors, possibly of cellular origin, may contribute to the T-cell depletion observed in AIDS.

A. HIV Genome and Life Cycle

To understand how HIV wrestles control of cellular metabolism, we need to review some aspects of its genome. In addition to the structural genes, the genome of HIV contains several regulatory genes which are important in controlling its expression and replication [for a review see Wong-Staal(1991)]. Of these, the Tat (trans-acting transcriptional activator) and Rev proteins are essential for virus replication. The *tat* gene is encoded by two exons, one coding for the first 72 amino acids and the second for the carboxy-terminal 14 amino acids. The predominant form of Tat found in infected cells is 86 amino acids. Depending on the splicing pattern, other species of unknown biological relevance have been identified. The N-terminal 72 amino acids, encoded by the first exon, appear to be sufficient for Tat activity as a transcriptional activator (Cullen, 1986). In contrast, full-length Tat appears to be important for regulation of a variety of cellular genes. Three structural domains have been proposed: an amino-terminal portion that is thought to assume an α-helix conformation, a central domain that contains a cluster of seven cysteine residues and a third domain composed of a stretch of basic amino acids. Most of the cysteines in the second domain of Tat, (with the exception of cys_{31}) are essential for its activity (Sadaie et al., 1990). Frankel et al. (1988) proposed that this region of Tat forms a metal-linked dimer with metal ions such as zinc and cadmium bridging cysteine-rich regions from each monomer. Later studies, however, have suggested that mutations of these cysteines have no effect on transcription. By constructing a genetic fusion between the amino-terminal DNA binding domain of the lambda repressor (as a reporter for dimerization) and Tat, Battaglia and colleagues (1994) showed that Tat has the ability to form dimers *in vivo*. Interestingly, site-directed mutagenesis suggested that cys_{37} was essential for this dimerization, invoking the possibility that metals may be involved in the interaction. The existence of a Jurkat protein, identified as an endopeptidase that inhibits this dimerization is also suggested (Battaglia et al., 1994).

B. Regulation of HIV Expression and the Function of Tat

Tat acts by binding to a sequence, termed the trans-activation response element (TAR), located downstream from the site of transcriptional initiation in the viral long terminal repeat (LTR) (Marciniak et al., 1990). Transcripts containing the TAR RNA sequence adopt a stable stem-loop structure (Muesing et al., 1987). The minimal TAR region needed for Tat binding includes sequences between nucleotides +19 and +44 from the start site of transcription (defined as +1) (Garcia et al., 1988). The association between Tat and TAR is very tight and specific, with a K_d of approximately 1×10^{-9} M (Muller et al., 1990). Because inhibitors of transcription also inhibit Tat function (Gentz et al., 1989) and because of its location, TAR is thought to function as an RNA, allowing Tat to be correctly positioned for transcriptional activation (Selby and Peterlin, 1990). Nevertheless, the exact mechanism of action of Tat is still controversial. Some investigators have suggested that Tat acts by increasing the rate of viral transcription rather than by modulating mRNA stability (Laspia et al., 1989) but others (Drysdale and Pavlakis, 1991) have proposed that Tat acts at both the transcriptional and post-transcriptional levels. Kessler and Mathews (1992) suggested that Tat may act as an anti-terminator, enabling transcription complexes to overcome putative pause sites. The Tat inhibitor Ro5-3335 has been shown to inhibit both transcriptional initiation as well as elongation from the HIV LTR, suggesting that there may be a common mechanism for Tat-mediated transcription and elongation (Cupelli and Hsu, 1995). In a series of elegant experiments, the rate-limiting step in trans-activation was shown to be the initial contact between Tat, TAR and the promoter region in the LTR, but not pre-initiation transcriptional complex formation (Jeang and Berkhout, 1992). On the other hand, because of its ability to functionally substitute for the Herpes simplex virus-derived VP16 trans-activator protein, Tat may act as a classical acidic transcriptional factor. These activators are important in activated regulation rather than in basal regulation, and interact with coactivators and not with basal transcriptional factors (Tiley et al., 1992). Tat may interact with additional cellular proteins, possibly bringing these in the vicinity of the transcriptional pre-initiation complex (Leguern et al., 1993). In fact, Tat has been shown to cooperate effectively with the cellular transcription factor Sp1 (Southgate and Green, 1991). The need for specific co-activators plus the fact that a protein coded for by in human chromosome 12 is required for optimal Tat activity (Alonso et al., 1992) provide a clue for the host tropism of HIV. Furthermore, levels of *trans*-activation of the HIV LTR by Tat show marked species differences.

The viral LTR contains a variety of *cis*-acting sequence elements other than TAR that are thought to interact directly with cellular transcriptional factors (such as AP-1, Sp1, and NFkB). Even though Tat-dependent transcription from the viral LTR occurs in the absence of these sequence elements (Doppler et al., 1992), the presence of some of these cellular factors increases Tat-dependent transcription (Liu et al., 1992). In glial cell lines, Tat-dependent activation of the HIV LTR occurred in a TAR-independent fashion, relying solely on the NF-κB domain of the LTR (Taylor et al., 1992b). The authors concluded that upstream promoter elements within the LTR could mediate Tat trans-activation. Additional cellular factors which bind to the TAR element and stimulate Tat-dependent transcription from the viral LTR have been identified (Gaynor, 1992; Marciniak et al., 1990; Wu et al., 1991).

C. Tat as a Regulator of Cellular Gene Expression

In addition to its transcriptional activity on HIV genes, Tat may also influence cellular phenotype by affecting the expression of cellular genes. In most cases, the two-exon Tat has been shown to be the culprit. For example, Tat expression led to transformation of primary keratinocyte cultures (Kim et al., 1992), increased collagen expression in glioblastoma cells (Taylor et al., 1992a), decreased IL-2 receptor expression in Jurkat cells (Purvis et al., 1992), TNF induction in a human lymphoblastoid cell line (Sastry et al., 1990), and repressed Mn-SOD in HeLa cells (Flores et al., 1993) (discussed later). The tumorigenic potential of Tat was demonstrated by the development of dermal lesions resembling Kaposi's sarcoma (KS) in mice transgenic for this protein (Vogel et al., 1988), by alterations in the expression of putative angiogenesis factors after introduction of Tat-producing cells into nude mice (Huber et al., 1992) and by the growth promotion of AIDS-KS cells exposed to Tat-containing supernates (Ensoli et al., 1990).

The major histocompatibility antigen expression was transiently decreased in HIV-1 infected T cell lines (Scheppler et al., 1989). This decrease could be reproduced in HeLa cells transfected with the two-exon *tat* gene (Howcroft et al., 1993), suggesting that the level of Tat expression in *tat*-transfected cells is functionally comparable to that of HIV-infected cells. Furthermore, these experiments define a function for the two-exon Tat that is distinct from the one-exon Tat and indicate that the HIV-1 Tat protein can act both as a transcriptional repressor or an activator. Tat can also stimulate the production of immunoglobulin and interleukin 6 by uninfected peripheral blood mononuclear cells in a dose-dependent manner at doses as small as 1 ng/ml. This effect was abrogated by Tat monoclonal antibodies and by 2,3-dimercapto-1-propanol, a reducing agent (Rautonen et al., 1994). Kubota et al. (1990) showed that this drug, used to treat heavy metal-poisoning, inhibited trans-activation by the Tat protein and also interfered with virus production.

An RGD (arg-gly-asp) sequence in one of Tat's domains is thought to interact with cellular integrins and stimulate the uptake of this viral protein (Weeks et al., 1993; Barillari et al., 1993). Once inside the target cells, Tat translocates into the nucleus where it may transcellularly transactivate HIV-LTR-mediated reporter gene expression (Frankel and Pabo, 1988). Co-cultivation experiments demonstrated that Jurkat-*tat* cells trans-activated the HIV-LTR in HeLa cells through cell-to-cell contact (Helland et al., 1991). Neuronal cells can effectively take up recombinant Tat from the surrounding medium (Kolson et al., 1994), a phenomenon that may explain some of the AIDS-associated neurological dysfunctions, even in the absence of productive infection. In fact, Tat activates excitatory non-N-methyl-D-aspartate amino acid receptors which results in neurotoxicity (Magnuson et al., 1995). Recent evidence suggests that the worst immunodeficiencies are observed when a macrophage trophic-strain of virus is used (Mosier et al., 1993). A clonal form of HIV which was associated with the early stages of Kaposi's sarcoma (Mcgrath et al., 1995), was localized to the tumor-associated macrophages. The infected macrophages coexpressed high levels of Tat, IL-6 and basic fibroblast growth factor, suggesting that these cells were providing the necessary environment for the growth of the tumor. If the oxidative balance of neurons is altered by Tat (see below), an increased level of reactive oxygen species would act synergistically with the increased growth factors to induce proliferation of the target cells and the appearance of a tumor. Through Tat, HIV infection affects cellular functions of uninfected cells, possibly contributing to some of the AIDS associated pathologies. The ability of this protein to exert the effects of the infection at a distance suggests that not only is the viral load a problem but also that there is an amplification of the pathogenic effect of the virus with the expected consequences to the organism.

Ultraviolet radiation induces HIV gene expression in transgenic mice carrying both the viral LTR and the *tat* gene (Frucht et al., 1991; Stanley et al., 1989). One proposed mechanism to explain these observations is that UV-induced cellular damage stimulates DNA repair functions which in turn induce transcription of viral DNA, expression of cellular proteins, and accelerated viral production (Radman, 1980). When HeLa or primary fibroblast cells were irradiated with UV light, HIV-1, collagenase, c-*Fos*, and metallothionein were induced (Stein et al., 1989). Radiation could damage DNA either by direct interaction with the DNA, or by the formation of active oxygen species, which also damage DNA. Cytokines, including TNF, stimulate HIV transcription and replication (Roederer et al., 1990). Stimulation of cell lines with these cytokines results in the production of oxidants and consumption of intracellular glutathione (Mihm et al., 1991). Hydrogen peroxide stimulates HIV transcription in a Jurkat subclone (Schreck et al., 1991), and results in glutathione consumption (Mihm et al., 1991). These activators may be inhibited by N-acetyl cysteine, a cysteine analog that replenishes intracellular GSH (Roederer et al., 1990; Kalebic et al., 1991). The idea that oxidants are the common pathways involved in HIV and transcriptional factor activation is further validated by the fact that antioxidants reverse the above effects. The stimuli that lead to transcriptional activation of certain genes, or even cellular proliferation, are the same stimuli which lead to HIV activation (Virelizier, 1990). Conversely, hindering the activation of transcriptional factors by antioxidants should also block viral activation. For example, antioxidants inhibit the activation of NF-κB by the viral transactivator *tax* from human T-cell leukemia virus type I in Jurkat cells (Frucht et al., 1991). Peroxyl radical scavenging blocks activation of both NF-κB and of HIV-1 LTR-driven gene expression (Israel et al., 1992). These data suggest that increased oxidative stress is a prerequisite for full activation of the transduction pathways regulating the activity of transcriptional factors and/or virus replication. Interestingly, patients who have AIDS (Staal et al., 1992) or even those who are HIV-positive but pre-symptomatic have decreased plasma and cellular glutathione levels and lower vitamin E levels (Javier et al., 1992). Exogenous reducing agents such as N-acetyl cysteine, glutathione, and diethyldithiocarbamate *suppress* HIV expression in chronically infected monocytes (Kalebic et al., 1991). Clinically, glutathione in conjunction with AZT treatment led to recovery of plasma thiol levels (Eck et al., 1989).

IV. MnSOD REPRESSION AND OXIDATIVE STRESS

Wong et al. (1991) demonstrated that treatment of the HIV-infected T-cell line HUT-78 with TNF failed to induce Mn-SOD. Moreover, the infected cells had increased sensitivity to heat and radiation, possibly because of reduced expression of Mn-SOD. Surprisingly, the unstimulated basal level of MnSOD was also lower in the HIV-infected cells. Our own work suggests a possible explanation for these findings: HeLa cells which produce the HIV regulatory protein Tat have depressed Mn-SOD levels. While the cytosolic Cu,Zn-SOD isozyme levels are unchanged (compared to parental HeLa), the Mn-SOD isozyme expression is repressed 50% ($P < 0.001$). In addition, levels of glucose-6-phosphate dehydrogenase (G-6-PD) in HeLa-*tat* cells are only about 10% of parental. Both the activity and steady-state levels of Mn-SOD-specific RNAs are much lower in HeLa-*tat* cells than in the parental cells. In contrast, the levels of G-6-PD RNA appear to be the same in both cell types. Clearly, different mechanisms are involved in the repression of these two antioxidant enzyme activities. Nevertheless, the end result of repression of these two enzymes is a generalized state of oxidative stress. Carbonyl proteins (an index of lipid peroxidation and protein oxidation) were 75% higher while total sulfhydryl content was 32% lower in HeLa-*tat* cells when compared to the parental cell line, reflecting a substantial level of oxidative stress and a drain on the cell's important antioxidant compound glutathione. In fact, when assayed for acid-soluble reduced glutathione, HeLa cells contained $3.23 \pm 0.87 \mu g/mg$ protein, while HeLa-*tat* cells contained $1.64 \pm 0.11 \mu g/mg$. This mechanistic link between lipid peroxidation and sulfhydryl consumption is discussed later but may account for the decreased glutathione levels observed clinically in HIV-infected patients (Buhl et al., 1989; Eck et al., 1989) and suggest that rather than a response to the infection, these decreases may be a direct result of Mn-SOD and/or G-6-PD repression. Recently, Westendorp et al. (1995) confirmed our results by showing that HeLa cells independently established to produce the HIV-1 Tat protein had repressed expression of Mn-SOD. Cu,Zn-SOD activity was not affected by Tat. The decreased Mn-SOD was also associated with decreased glutathione levels as well as a lower ratio of reduced:oxidized glutathione. Similar findings were reported for Jurkat cells treated with exogenously added Tat. Only the 2-exon Tat protein had this effect, which is consistent with the Tat sequence in our HeLa-*tat* cell culture system.

How is it that Tat has such effects on MnSOD expression? The ability of Tat to affect HIV gene expression results from binding to the TAR RNA stem–loop structures. Computer modeling of the Mn-SOD transcript predicts a stable stem loop structure whose three-dimensional configuration is very homologous to the TAR stem–loop structure. We sought evidence for this binding and found that indeed there was a physical interaction between Tat and the MnSOD transcripts, which could account for the down-regulation of this enzyme. Furthermore, Tat inhibitors such as Ro5-3335 diminished this association, corroborating its specificity. A recent report by Buonaguro et al. (1994) concluded that Tat-specific induction of TNF-β depended on the presence of a predicted TAR-like stem loop in the TNF-β mRNA leader region. These experiments suggest that Tat has the ability to bind to specific cellular sequences and consequently affect the expression of those genes.

V. CONSEQUENCES OF OXIDATIVE STRESS

The realization that, as part of its cytopathic effect, HIV infection shifts the cellular redox status by blocking the cellular defenses precisely designed to circumvent oxidative damage was surprising. Ultimately, a cell thus challenged will die, a condition which, at first glance, may appear evolutionarily disadvantageous for the virus. Nevertheless, exposure to oxidants challenges cellular systems which respond in many cases by activation of transcriptional factors and altered gene expression, conditions that may favor virus replication. One of the factors activated by changes in redox status is NF-κB (Toledano and Leonard, 1991). This factor also modulates HIV-1 gene expression (Gaynor, 1992) through the NF-κB enhancer sequence present in the HIV-1 LTR.

Another transcriptional factor modulated by the redox status of the cell is AP-1, a heterodimer composed of Fos and Jun proteins (Nose et al., 1991). The induction of these *onc*-genes may involve the action of protein kinase C (PK-C), and can be partly abrogated by N-acetyl cysteine (Datta et al., 1992).

Besides transcriptional activation, oxidants also affect cell growth, possibly through the activation of *onc*-genes which are responsible for cell cycle control and differentiation. Exposure of cells to either exogenous oxidants or oxidants generated by ischemia/reperfusion (Brand et al., 1992) leads to the induction of c-*jun*, c-*fos* (Devary et al., 1991; Nose et al., 1991) or c-*myc* (Crawford et al., 1988), increased cellular proliferation (Rao and Berk, 1992), and lung hyperplasia and fibrosis in hamsters (Nakashima et al., 1991). The induction of c-*myc* is very often accompanied by induction of p53, which in turn accelerates apoptosis, or programmed cell death.

To counter the effects of free radicals on cell proliferation, Mn-SOD expression appears to favor a differentiated state: activity is increased in differentiating trophoblasts (DellaCorte and Stirpe, 1972) and in growth-arrested endothelial cells. Friend erythroleukemia cells that are induced to differentiate increase their Mn-SOD activity (Beckman et al., 1989), a response also observed when liposome-encapsulated Mn-SOD is added to the culture medium. Cu,Zn-SOD reverses the tumor promoting effect of 12-o-tetradecanoyl phorbol-13 acetate (TPA) and of superoxide exposure in mouse epidermal cells (Nakamura et al., 1988). In a series of elegant experiments, Church et al. (1993) demonstrated that overexpression of Cu,Zn-SOD reverses the malignant phenotype of melanoma cells. Thus, inhibition of the SOD enzymatic activities could potentially cause neoplastic transformation. While humans have three SOD genes, only the Mn-SOD gene is known to be inducible in adult somatic tissues, and may account for more than 70% of the cell's total SOD activity (McCord et al., 1977). In fact, some investigators have suggested that the reason humans live longer is due to increased Mn-SOD levels compared to other mammals. Several studies have also suggested that the mRNA for Mn-SOD responds more readily than Cu,Zn-SOD to changes in differentiation state (Galeotti et al., 1989).

HIV-1 infection is known to activate Epstein Barr Virus (EBV) and c-*myc* in human B-lymphocytes (Astrin and Laurence, 1992), resulting in immortalization, but not apoptosis of these cells (Laurence and Astrin, 1991). Interestingly, one of the effects of Tat is transformation of cultured keratinocytes (Kim et al., 1992). Radiation, another exogenous source of free radicals, induces c-*fos*, an effect which is inhibited by the reductant N-acetyl cysteine (Datta et al., 1992).

The conclusion is that by inhibiting MnSOD, HIV infection leads to an accumulation of the superoxide radical, mobilization of iron from tissue ferritin (Biemond et al., 1984) or from proteins containing iron chromophores (Gardner and Fridovich, 1992; see Figure 1). The liberated iron catalyzes lipid peroxidation (Thomas et al., 1985), seriously affecting cell membrane structural integrity and function. The accumulating phospholipid hydroperoxides are reduced by phospholipid hydroperoxide glutathione peroxidase (Ursini et al., 1985), or the peroxidized fatty acids are liberated by phospholipases and subsequently reduced by glutathione peroxidase. These enzymes deplete GSH, and cytosolic NADPH is in turn drained by glutathione reductase in an attempt to maintain reduced glutathione levels. Because of reduced G-6-PD activity, the cell is unable to regenerate NADPH effectively, glutathione levels are compromised and a state of generalized oxidative stress results, increasing cellular susceptibility to further insults. To investigate whether oxidative stress may be exacerbated by other types of stress, we examined the ther-

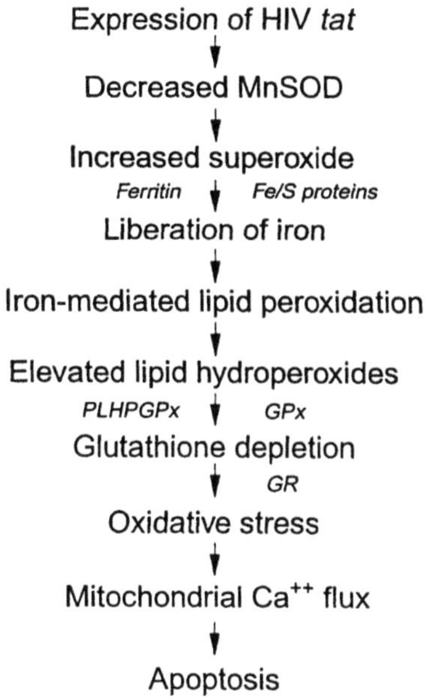

Figure 1. The sequence of events leading from expression of the Tat protein to the creation of oxidative stress, and ultimately to an increased incidence of cell death by apoptosis.

mal tolerances of HeLa and HeLa-*tat* cells. The parental cells were fully viable after 3 h at 44° C. The oxidatively-stressed Tat-producing cells were only 20% viable after the same thermal exposure, suggesting that their oxidative stress and thermal stress were additive. The thermal resistance of Tat-producing cells was increased by growth in the presence of 1 mM mercaptopropionylglycine (MPG) or 3 μM of the Tat-antagonist Ro5-3335. Other investigators have shown that hyperthermia selectively reduces the survival of SV40-transformed cells (with lower levels of SOD, catalase and GPX) while having relatively little effect on normal parental cells (Omar et al., 1987). TNF-mediated cytotoxicity for both HeLa and Jurkat T cells was enhanced in the presence of Tat (Westendorp et al., 1995), suggesting that a Tat-dependent alteration of the cellular redox status sensitizes these cells to TNF.

The observed Tat-induced increases in collagen synthesis (Taylor et al., 1992a) or cellular transformation (Kim et al., 1992) may be secondary to the increased concentration of reactive oxygen species. In fact, collagen induction by Tat-producing cells was suggested to occur though an indirect mechanism (Taylor et al., 1992a). Therefore, Tat-mediated repression of Mn-SOD can have an effect on the expression of other genes, possibly mediated by increased oxidative stress.

These biochemical findings may explain the numerous clinical observations of reduced vitamin E, glutathione, selenium, and other antioxidants observed at the various stages of symptomatic or asymptomatic HIV-1 infection (Buhl, 1994; Dworkin, 1994; Staal et al., 1992; Favier et al., 1994; Wang and Watson, 1994). More significantly, increased oxidative stress may predispose the affected cell and its uninfected neighbors to apoptosis and depletion, ultimately leading to AIDS.

VI. APOPTOSIS AND OXIDATIVE STRESS

Abnormalities of the immune system can be demonstrated early in HIV infection even though only a small number of T cells are infected. While normal lymphocytes respond to a mitogenic stimulus by proliferating, approximately 40% of the peripheral lymphocytes from HIV-infected patients undergo apoptosis instead. This behavior is puzzling given the fact that only 0.1% of these are infected with the virus. The progressive loss of T lymphocytes by this mechanism has been proposed to lead to the compromised immunity of AIDS (Ameisen and Capron, 1991; Ameisen, 1992). How does one account for this phenomenon *in vivo*, where some systemic factor must be affecting the behavior of uninfected cells? Because of its ability to be secreted and taken up by other cells, we believe that the HIV regulatory protein Tat is such a factor. When cells are exposed to this HIV protein, MnSOD is repressed and oxidative stress increases, raising their susceptibility to additional insults (see above).

At first it may appear that shifting the oxidative balance in favor of oxidants is always bad. In fact, mild oxidative stress induces normal cells to divide (Nishioka and Welsh, 1994; Burdon et al., 1994) and within certain limits is a normal regulator of cell function. When lymphocytes are stimulated by mitogen, their oxidative stress increases and their own production of oxidants increases as well. If, however, the rate of oxidant production exceeds the cellular capacity to detoxify them, then a series of events is triggered that may culminate in apoptosis. In a cell already oxidatively stressed by Tat, the combined effect of mitogen stimulation may produce apoptosis rather than proliferation. So how does one prevent apoptosis of these lymphocytes? The protein Bcl-2 (a product of the B-cell leukemia/lymphoma gene) inhibits most types of apoptotic cell death (Sentman et al., 1991), and it has been suggested to act via an antioxidant mechanism (Hockenbery et al., 1993). Others, however, have suggested that it may function as a prooxidant because of its ability to increase catalase levels in *E. coli* (Steinman, 1995). Its biochemical function is, however, still not clearly defined. Other antioxidants, such as thiol compounds can inhibit activation-induced death of T-cell hybridomas (Sandstrom et al., 1994a). The human T-cell leukemia virus Tax protein, which is an analogue of Tat, induces apoptosis in Rat-1 cells which is prevented by Bcl-2 (Yamada et al., 1994). On the other hand, endogenous Bcl-2 is incapable of inhibiting apoptosis of HIV-infected cells (Antoni et al., 1995). WR-1065, an antioxidant, protects thymocytes from apoptosis by inhibiting the endonuclease that is responsible for the DNA fragmentation observed (Ramakrishnan and Catravas, 1992). Treatment of normal fibroblasts with transforming growth factor-β results in the production of a substance that eliminates transformed fibroblasts via apoptosis (Jurgensmeier et al., 1994). Even though the nature of this substance was not determined, its apoptotic activity was abrogated by treatment with antioxidants and radical scavengers, suggesting that some product of increased oxidative stress (e.g., lipid peroxidation), may be the culprit. Interestingly, overexpression of Bcl-2 in a neural cell line prevents lipid peroxidation (Kane et al., 1993). By overexpressing the apoptosis-inhibiting adenovirus E1B protein in HIV-infected lymphocytes, some investigators demonstrated that these cells continuously produced HIV (Antoni et al., 1995), leading them to conclude that preventing apoptosis could be dangerous in this context. However, if the alternative is to lose a large proportion of the lymphocytes with the consequent immunodeficiency, then it would be preferable to live with a chronic, persistent infection while lymphocyte viability is preserved. Continued low levels of viral replication are not necessarily associated with disease progression if controlled over time (Pantaleo et al., 1995). If apoptosis is not prevented in the HIV-infected individuals, the result is a progressive loss of T-cells and ultimately AIDS.

It is obvious from the present discussion that increased oxidative stress may push the cell through two different pathways: proliferation or apoptosis. Whether one or the other is chosen probably depends on the proliferative state of the target cell, which in turn will affect the redox status. In addition, a cell may or may not be able to erect or bring down barriers or checkpoints to either pathway. These barriers may include the retinoblastoma gene product, p53, Bcl-2, c-*myc,* and even SOD. For example, in a normal physiological setting, the effect of cytotoxic T-lymphocytes (CTL) depends on the proliferative status of the target: log phase growth, transformation with c-*myc*, or infection of quiescent G_0 targets with herpes simplex virus-1, which induces a competent state of DNA synthesis, all increase target cell susceptibility to CTL-induced DNA fragmentation (Nishioka and Welsh, 1994). The relationship between the cell cycle and apoptosis has led Rubin et al. (1995) to propose that "abortive entry into the cell cycle may be responsible for the programmed death" of a variety of cell types. Persistently elevated levels of the $p34^{cdc2}$-kinase and cyclin B-associated histone H1 kinase have been found in T-cells undergoing activaton-induced cell death (Fotedar et al., 1995). The *onc*-gene c-*myc* also has a critical role in cell-cycle control as well as apoptosis (Green et al., 1994; Steller, 1995), and in some cell types participates in conjunction with cyclin A in the induction of this process (Hoang et al., 1994). Under normal cell culture conditions, overexpression of c-*myc* causes cells to proliferate. If growth factors are removed, certain c-*myc* overexpressors will undergo apoptosis instead (Askew et al., 1991; Evan et al., 1992).

Bcl-2 is another cellular checkpoint whose mission is to prevent apoptosis when the latter is inappropriately initiated. This protein is found in mitotic nuclei and its expression varies within the cell cycle (Lu et al., 1994). The observation that Bcl-2-inhibitable apoptosis can be demonstrated in anucleated cytoplasts, has prompted Jacobson et al. (1994) to propose that apoptosis, like the cell cycle, may be initiated by membrane events and orchestrated by a cytoplasmic regulator that has multiple intracellular targets. These investigators favor the possibility that RNA and protein synthesis are required for the regulation of the death program rather than for the synthesis of the components of the program itself. Recent work provides evidence for a membrane pathway to apoptosis that may be independent of nuclear events: TNF binding to bovine aortic endothelial cells leads to sphingomyelin hydrolysis, ceramide release and activation of a serine/threonine protein kinase which culminates in apoptosis of the target cell (Haimovitz-Friedman et al., 1994). Treatment with ionizing radiation had the same effect. Protein kinase C activation blocked both radiation-induced ceramide release and apoptosis, which could be restored by the addition of ceramide. The effects of ionizing radiation occurred even in the absence of a nucleus, indicating that direct DNA damage is not required for radiation-induced killing. PKC activation can also repress cell death that results from IL-2 withdrawal, but not prevent glucorticoid-induced cell death of CTL (Walker et al., 1993). In contrast, other reports have suggested that PKC mediates radiation-induced apoptosis of thymocytes (Ojeda et al., 1992). In addition, TPA induces growth arrest and differentiation of HL-60 promyelocytes via activation of the PKC-β isozyme. These cells eventually adhere and undergo apoptosis. Thus, the effects of PKC on apoptosis are conflicting and may depend on the particular PKC isozyme activated in a specific cell type.

The product of the tumor suppressor gene p53 has been implicated in DNA-damage-induced growth arrest and apoptosis. On the other hand, if c-*myc* expression is also increased in these damaged cells, it apparently binds to p53, preventing apoptosis and instead leading to cell proliferation and eventual transformation (Shi et al., 1992). Interestingly, increased oxidative stress induces c-*myc*, possibly allowing the cells to become competent at proliferation (Crawford et al., 1988).

Longo et al., (1995) using a model system to probe for protein-protein interactions suggested that the HIV Tat protein interacts with the human p53 protein. Such an interaction may either inactivate the p53 protein, leading to transformation, or may promote the formation of p53 dimers, favoring apoptosis.

We believe that depending on the differentiation or proliferation state every cell type will exhibit a characteristic position on the superimposed curves shown on Figure 2. The two vertical lines represent barriers, either genetic or environmental, to either proliferation or death. Shifting these barriers in the direction of reductants or oxidants will result in a cell that will favor quiescence, proliferation or death. If the cell is quiescent or terminally differentiated, then its initial position will be on the point labeled A. Because of a mitotic block, these cells have such constraints on the ability to proliferate that when the concentration of oxidants increases sufficiently, their response will be to die by either apoptosis or necrosis. On the other hand, cells represented by curve B still maintain a genetic program that allows them to proliferate when subjected to a stimulus which may be an antigen, a mitogen, increased oxidants, or a combination of these. The initial stimulus may provide the "priming" event that lowers the initial barriers (i.e., G_0/G_1 transition), but an in-

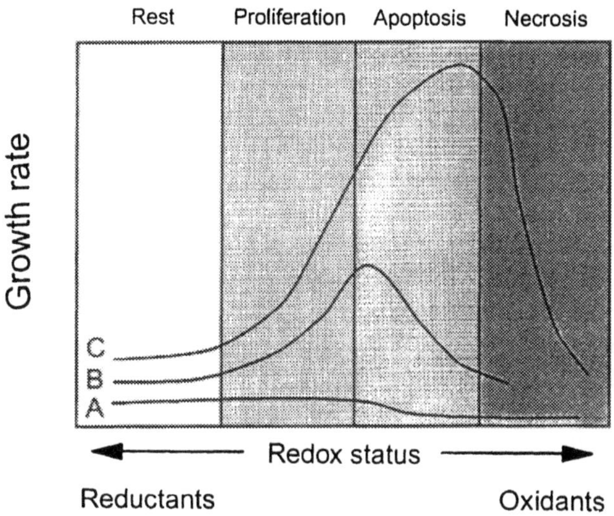

Figure 2. Hypothetical response of various cell types to increasing oxidative stress: (**A**) post-mitotic terminally differentiated cells, (**B**) cells capable of proliferation such as fibroblasts or lymphocytes, (**C**) transformed cells.

crease in oxidative stress provides the driving force for continuous proliferation. Such a model predicts that mitogenic stimulation of lymphocytes would be prevented in the presence of a highly reducing environment. If, on the other hand, the cell is continuously proliferating, as is the case with transformed cells, its initial position would be on curve C. These cells presumably have no barriers to proliferation, and a subcytotoxic dose of an oxidant may provide a further stimulus for proliferation. An even higher concentration of the oxidant would then push the cell beyond the optimal point of division and into the apoptotic range. These superimposed curves illustrate the fact that biological responses to oxidative stress are not linear but rather bell-shaped. The observed effect of an antioxidant depends on the starting position on any one of these superimposed curves (whether the stimulation is sub-optimal or supraoptimal). For example, low doses of vitamin E may potentiate the immune response but higher dosages are immunosuppressive (Yasunaga et al., 1982). From a biochemical perspective, antioxidants are known to prevent the induction of ornithine decarboxylase in response to mitogens (Johnson, 1989). This enzyme is involved in the synthesis of polyamines whose role is thought to involve regulation of the cell cycle. Interestingly, the redox cycling agent 2,3-dimethoxy-1,4-naphthoquinone (DMNQ) stimulates growth, triggers apoptosis, or causes necrosis of pancreatic insulinoma cells, depending on the dose and duration of the exposure (Dypbukt et al., 1994). The outcome of this treatment was dependent upon the activity of ornithine decarboxylase and consequent polyamine synthesis, suggesting a relationship to the cell cycle and/or proliferative state of the cell.

By using this curve, we can predict the immunological outcome of a variety of diseases which have been noted to result from alterations in antioxidant enzymes. For example, the immunodeficiency observed in Down's syndrome may result from the inability of lymphocytes to undergo blast transformation and clonal expansion. The elevated levels of Cu,Zn-SOD alter the balance and maintain the cell in a reduced state. A stimulus that would otherwise result in proliferation simply puts the cell in a more oxidized state (a position further to the right of the curve). In HIV infection and familial ALS (Lou Gehrig's disease), Mn-SOD and Cu,Zn-SOD are deficient, respectively. If affected lymphocytes are in an oxidative state due to SOD deficiency, the additional stress caused by mitogen stimulation may push them into the apoptotic range. The result is lymphopenia which is mild in ALS but severe in AIDS.

For a cell that is continuously exposed to Tat and whose oxidative status has been altered by this viral protein, the outcome of such an interaction may depend on its position on the hypothetical curve. For example, under conditions of serum starvation, Tat-positive PC12 clones showed a significantly higher proliferation rate with respect to mock-transfected cells or cells producing lower levels of Tat (Milani et al., 1993). Furthermore, Tat stimulates the growth of cells derived from Kaposi's sarcoma lesions of AIDS patients (Ensoli et al., 1990). On the other hand, higher concentrations of Tat have been shown to be cytotoxic (Milani et al., 1993). Synthetic Tat peptides inhibit the proliferative responses of CD4+ antigen-specific T-cell clones (Chirmule et al., 1995). The observed effects were dose-dependent and stimulus-specific. Interestingly, in co-culture experiments, soluble Tat protein secreted by HeLa cells transfected with the *tat* gene also inhibited the proliferation of lymphocytes (Chirmule et al., 1995). In addition, antigen-induced lymphocyte proliferation is inhibited by this protein (Viscidi et al., 1989). The majority of these studies used [^{3}H]-thymidine incorporation as the end-point for proliferation. Thus, failure to proliferate because of increased cell death would have gone undetected. In fact, Tat has been shown to increase apoptosis and death of uninfected Jurkat cells (Li et al., 1995; Purvis et al., 1995). On the other hand, expression of the *tat* gene prevented apoptosis of several cell lines after serum deprivation (Zauli et al., 1993) and to a variety of other stimuli, including HIV infection (Gibellini et al., 1995). These variable effects may in fact depend on the proliferative state of the cell under study.

CONCLUSIONS

Because antioxidants inhibit both apoptosis and transformation, it is tempting to speculate that there is a common pathway which is sensitive to host redox status, and that depending on the barriers present, may lead in either direction. If, in fact, AIDS results when oxidatively-stressed T cell populations are eliminated by apoptosis, interventions aimed at restoring normal redox balance would have great value in clinical application. Nevertheless, because of the bell-shaped nature of the cellular response to oxidative stress, attempts to blindly modify a patient's oxidative status by administration of arbitrary amounts of reducing agents such as N-acetyl cysteine are producing understandably mixed results. Therefore, dosages of antioxidants should be carefully monitored, possibly using lymphocyte functional tests as an end-point.

REFERENCES

Akaike, T., & M., Oda, T., Doi, T., Ijiri, S., Araki, S., & Maeda, H. (1990). Dependence on $O_2^{\bullet -}$ generation by xanthine oxidase of pathogenesis of influenza virus infection in mice. J. Clin. Invest. 85, 739–745.

Alonso, A., Derse, D., & Peterlin, B.M. (1992). Human chromosome-12 is required for optimal interactions between Tat and TAR of human immunodeficiency virus type-1 in rodent cells. J. Virol., 66, 4617–4621.

Ameisen, J.C. (1992). Programmed cell death and AIDS: From hypothesis to experiment. Immunol. Today 13, 388–391.

Ameisen, J.C. (1994). Programmed cell death (apoptosis) and AIDS pathogenesis. AIDS Res. Hum. Retroviruses, 10, S3–S5.

Ameisen, J.C., & Capron, A. (1991). Cell dysfunction and depletion in AIDS: the programmed cell death hypothesis. Immunol. Today 12, 102–105.

Antoni, B.A., Sabbatini, P., Rabson, A.B., & White, E. (1995). Inhibition of apoptosis in human immunodeficiency virus-infected cells enhances virus production and facilitates persistent infection. J. Virol. 69, 2384–2392.

Arora, D.J.S., & Henrichon, M. (1994). Superoxide anion production in influenza protein-activated NADPH oxidase of human polymorphonuclear leukocytes. J. Infect. Dis. 169, 1129–1133.

Askew, D.S., Ashmun, R.A., Simmons, B.C., & Cleveland, J.L. (1991). Constitutive *c-myc* expression in an IL-3 dependent myeloid cell line suppresses cell cycle arrest and accelerates apoptosis. Oncogene 6, 1915–1922.

Astrin, S.M. and Laurence, J. (1992). Human immunodeficiency virus activates *c-myc* and Epstein-Barr virus in human B lymphocytes. Ann. N Y Acad. Sci. 651, 422–432.

Babior, B.M. (1978). Oxygen-dependent microbial killing by phagocytes. N. Eng. J. Med. 298, 659–668 and 721–725.

Barillari, G., Gendelman, R., Gallo, R.C., & Ensoli, B. (1993). The Tat protein of human immunodeficiency virus type-1, a growth factor for AIDS Kaposis sarcoma and cytokine-activated vascular cells, induces adhesion of the same cell types by using integrin receptors recognizing the RGD amino acid sequence. Proc. Natl. Acad. Sci. USA 90, 7941–7945.

Battaglia, P.A., Longo, F., Ciotta, C., Delgrosso, M.F., Ambrosini, E., & Gigliani, F. (1994). Genetic tests to reveal Tat homodimer formation and select Tat homodimer inhibitor. Biochem. Biophys. Res. Commun. 201, 701–708.

Beckman, B.S., Balin, A.K., & Allen, R.G. (1989). Superoxide dismutase induces differentiation of friend erythroleukemia cells. J. Cell. Physiol. 139, 370–376.

Biemond, P., van Eijk, H.G., Swaak, A.J.G., & Koster, J.F. (1984). Iron mobilization from ferritin by superoxide derived from stimulated polymorphonuclear leukocytes. Possible mechanism in inflammation diseases. J. Clin. Invest. 73, 1576–1579.

Boyes, B.E., Walker, D.G., McGeer, E.G., & O'Kusky, J.R. (1989). Increased uric acid in the developing brain and spinal cord following cytomegalovirus infection. J. Neurochem. 53, 1719–1723.

Brand, T., Sharma, H.S., Fleischmann, K.E., Duncker, D.J., Mcfalls, E.O., Verdouw, P.D., & Schaper, W. (1992). Proto-oncogene expression in porcine myocardium subjected to ischemia and reperfusion. Circ. Res. 71, 1351–1360.

Bravard, A., Sabatier, L., Hoffschir, F., Ricoul, M., Luccioni, C., & Dutrillaux, B. (1992). *Sod2*—A new type of tumor-suppressor gene. Int. J. Cancer 51, 476–480.

Buhl, R., Jaffe, H.A., Holroyd, K.J., & et al., (1989). Systemic glutathione deficiency in symptom-free HIV-seropositive individuals. Lancet 2, 1294–1298.

Buhl, R. (1994). Imbalance between oxidants and antioxidants in the lungs of HIV-seropositive individuals. Chem. Biol. Interactions 91, 147–158.

Buonaguro, L., Buonaguro, F.M., Giraldo, G., & Ensoli, B. (1994). The human immunodeficiency virus type 1 Tat protein transactivates tumor necrosis factor β gene expression through a tar-like structure. J. Virol. 68, 2677–2682.

Burdon, R.H., Alliangana, D., & Gill, V. (1994). Endogenously generated active oxygen species and cellular glutathione levels in relation to BHK-21 cell proliferation. Free Radical Res. 21, 121–133.

Buttke, T.M., & Sandstrom, P.A. (1994). Oxidative stress as a mediator of apoptosis. Immunol. Today 15, 7–10.

Chirmule, N., Than, S., Khan, S.A., & Pahwa, S. (1995). Human immunodeficiency virus Tat induces functional unresponsiveness in T cells. J. Virol. 69, 492–498.

Church, S.L., Grant, J.W., Ridnour, L.A., Oberley, L.W., Swanson, P.E., Meltzer, P.S., & Trent, J.M. (1993). Increased manganese superoxide dismutase expression suppresses the malignant phenotype of human melanoma cells. Proc. Natl. Acad. Sci. USA 90, 3113–3117.

Crawford, D., Zbinden, I., Amstad, P., & Cerutti, P. (1988). Oxidant stress induces the proto-oncogenes c-*fos* and c-*myc* in mouse epidermal cells. Oncogene 3, 27–32.

Cullen, B.R. (1986). Transactivation of human immunodeficiency virus occurs via a bimodal mechanism. Cell 46, 973–982.

Cupelli, L.A., & Hsu, M.C. (1995). The human immunodeficiency virus type 1 Tat antagonist, Ro 5-3335, predominantly inhibits transcription initiation from the viral promoter. J. Virol. 69, 2640–2643.

Datta, R., Hallahan, D.E., Kharbanda, S.M., Rubin, E., Sherman, M.L., Huberman, E., Weichselbaum, R.R., & Kufe, D.W. (1992). Involvement of reactive oxygen intermediates in the induction of c-*jun* gene transcription by ionizing radiation. Biochemistry 31, 8300–8306.

DellaCorte, E., & Stirpe, F. (1972). The regulation of rat liver xanthine oxidase: Involvement of thiol groups in the conversion of the enzyme activity from dehydrogenase (type D) into oxidase (type O) and purification of the enzyme. Biochemistry 126, 739–745.

Deloria, L., Abbott, V., Gooderham, N., & Mannering, G.J. (1985). Induction of xanthine oxidase and depression of cytochrome P-450 by interferon inducers: Genetic difference in the responses of mice. Biochem. Biophys. Res. Commun. 131, 109–114.

Devary, Y., Gottlieb, R.A., Lau, L.F., & Karin, M. (1991). Rapid and preferential activation of the c-*jun* gene during the mammalian UV response. Mol. Cell Biol. 11, 2804–2811.

Doppler, C., Schalasta, G., Amtmann, E., & Sauer, G. (1992). Binding of NF-κB to the HIV-1 LTR is not sufficient to induce HIV-1 LTR activity. AIDS Res. Hum. Retroviruses 8, 245–252.

Drysdale, C.M., & Pavlakis, G.N. (1991). Rapid activation and subsequent down-regulation of the human immunodeficiency virus type 1 promoter in the presence of Tat: possible mechanisms contributing to latency. J. Virol. 65, 3044–3051.

Dworkin, B.M. (1994). Selenium deficiency in HIV infection and the acquired immunodeficiency syndrome (AIDS). Chem. Biol. Interact. 91, 181–186.

Dypbukt, J.M., Ankarcrona, M., Burkitt, M., Sjoholm, A., Strom, K., Orrenius, S., & Nicotera, P. (1994). Different prooxidant levels stimulate growth, trigger apoptosis, or produce necrosis of insulin-secreting RINm5F cells—The role of intracellular polyamines. J. Biol. Chem. 269, 30553–30560.

Eck, H.P., Gmunder, H., Hartmann, M., Petzoldt, D., Daniel, V., & Droge, W. (1989). Low concentrations of acid soluble thiol (cysteine) in blood plasmas of HIV-1-infected patients. Biol. Chem. Hoppe. Seyler. 370, 101–108.

Ensoli, B., Barillari, G., Salahuddin, Z., Gallo, R.C., & Wong-Staal, F. (1990). Tat protein of HIV-1 stimulates growth of cells derived from Kaposi's sarcoma lesions of AIDS patients. Nature 345, 84–86.

Evan, G.I., Wyllie, A.H., Gilbert, C.S., Littlewood, T.D., Land, H., Brooks, M., Waters, C.M., Penn, L.Z., & Hancock, D.C. (1992). Induction of apoptosis in fibroblasts by c-myc protein. Cell 69, 119–128.

Favier, A., Sappey, C., Leclerc, P., Faure, P., & Micoud, M. (1994). Antioxidant status and lipid peroxidation in patients infected with HIV. Chem. Biol. Interact. 91, 165–180.

Flores, S.C., Marecki, J.C., Harper, K.P., Bose, S.K., Nelson, S.K., & McCord, J.M. (1993). Tat protein of human immunodeficiency virus type 1 represses expression of manganese superoxide dismutase in HeLa cells. Proc. Natl. Acad. Sci. USA 90, 7632–7636.

Fotedar, R., Flatt, J., Gupta, S., Margolis, R.L., Fitzgerald, P., Messier, H., & Fotedar, A. (1995). Activation-induced T-cell death is cell cycle dependent and regulated by cyclin B. Mol. Cell Biol. 15, 932–942.

Frankel, A.D., Chen, L., Cotter, R.J., & Pabo, C.O. (1988). Dimerization of the Tat protein from human immunodeficiency virus: A cysteine-rich peptide mimics the normal metal-linked dimer interface. Proc. Natl. Acad. Sci. USA 85, 6297–6300.

Frankel, A.D., & Pabo, C.O. (1988). Cellular uptake of the Tat protein from human immunodeficiency virus. Cell 55, 1189–1193.

Frucht, D.M., Lamperth, L., Vicenzi, E., Belcher, J.H., & Martin, M.A. (1991). Ultraviolet radiation increases HIV-long terminal repeat-directed expression in transgenic mice. AIDS Res. Hum. Retroviruses 7, 729–733.

Fuchs, D., Gruber, A., Uberall, F., & Wachter, H. (1994). Oxidative stress and apoptosis. Immunol. Today 15, 496.

Galeotti, T., Wohlrab, H., Borrello, S., & De Leo, M.E. (1989). Messenger RNA for manganese and copper-zinc superoxide dismutases in hepatomas: Correlation with degree of differentiation. Biochem. Biophys. Res. Comm. 165, 581–589.

Garcia, J.A., Harrich, D., Pearson, L., Mitsuyasu, R., & Gaynor, R.B. (1988). Functional domains required for Tat-induced transcriptional activation of the HIV-1 long terminal repeat. EMBO J. 7, 3143–3147.

Gardner, P.R., & Fridovich, I. (1992). Inactivation–reactivation of aconitase in *Escherichia-coli*—a sensitive measure of superoxide radical. J. Biol. Chem. 267, 8757–8763.

Gaynor, R. (1992). Cellular transcription factors involved in the regulation of HIV-1 gene expression. AIDS 6, 347–363.

Gentz, R., Chen, C-H., & Rosen, C.A. (1989). Bioassay for trans-activation using purified human immunodeficiency virus Tat-encoded protein: Trans-activation requires mRNA synthesis. Proc. Natl. Acad. Sci. USA 86, 821–824.

Gibellini, D., Caputo, A., Celeghini, C., Bassini, A., Laplaca, M., Capitani, S., & Zauli, G. (1995). Tat-expressing Jurkat cells show an increased resistance to different apoptotic stimuli, including acute human immunodeficiency virus-type I (HIV-1) infection. Br. J. Haematol. 89, 24–33.

Green, D.R., Mahboubi, A., Nishioka, W., Oja, S., Echeverri, F., Shi, Y.F., Glynn, J., Yang, Y.L., Ashwell, J., & Bissonnette, R. (1994). Promotion and inhibition of activation-induced apoptosis in T-cell hybridomas by oncogenes and related signals. Immunol. Rev. 142321–342,

Groux, H., Torpier, G., Monte, D., Mouton, Y., Capron, A., & Ameisen, J.C. (1992). Activation-induced death by apoptosis in CD4+ T cells from human immunodeficiency virus-infected asymptomatic individuals. J. Exp. Med. 175, 331–340.

Haimovitz-Friedman, A., Kan, C.C., Ehleiter, D., Persaud, R.S., McLoughlin, M., Fuks, Z., & Kolesnick, R.N. (1994). Ionizing radiation acts on cellular membranes to generate ceramide and initiate apoptosis. J. Exp. Med. 180, 525–535.

Helland, D.E., Welles, J.L., Caputo, A., & Haseltine, W.A. (1991). Transcellular transactivation by the human immunodeficiency virus type 1 Tat protein. J. Virol. 65, 4547–4549.

Hennet, T., Peterhans, E., & Stocker, R. (1992). Alterations in antioxidant defences in lung and liver of mice infected with influenza-A virus. J. Gen. Virol. 73, 39–46.

Hoang, A.T., Cohen, K.J., Barrett, J.F., Bergstrom, D.A., & Dang, C.V. (1994). Participation of cyclin A in Myc-induced apoptosis. Proc. Natl. Acad. Sci. USA 91, 6875–6879.

Hockenbery, D.M., Oltvai, Z.N., Yin, X.M., Milliman, C.L., & Korsmeyer, S.J. (1993). Bcl-2 functions in an antioxidant pathway to prevent apoptosis. Cell 75, 241–251.

Howcroft, T.K., Strebel, K., Martin, M.A., & Singer, D.S. (1993). Repression of MHC class-I gene promoter activity by 2-exon *tat* of HIV. Science 260, 1320–1322.

Huber, B.E., Richards, C.A., Martin, J.L., & Wirth, P.J. (1992). Alterations in tumor angiogenesis associated with stable expression of the HIV *tat* gene. Mol. Carcinogen. 5, 293–300.

Ikeda, T., Shimokata, K., Daikoku, T., Fukatsu, T., Tsutsui, Y., & Nishiyama, Y. (1992). Pathogenesis of cytomegalovirus-associated pneumonitis in ICR mice—Possible involvement of superoxide radicals. Arch. Virol. 127, 11–24.

Israel, N., Gougerotpocidalo, M.A., Aillet, F., & Virelizier, J.L. (1992). Redox status of cells influences constitutive or induced NF-κB translocation and HIV long terminal repeat activity in human T-cell and monocytic cell lines. J. Immunol. 149, 3386–3393.

Jacobson, M.D., Burne, J.F., & Raff, M.C. (1994). Programmed cell death and Bcl-2 protection in the absence of a nucleus. EMBO J. 13, 1899–1910.

Javier, J.J., Todyce-Baum, M.K., Beach, R.S., Gavancho, M., Cabrejos, C., & Mantero-Atienza, E. (1992). Antioxident micronutrients and immune functions in HIV-1 infection. FASEB Proc. 1, 234–235.

Jeang, K.T., & Berkhout, B. (1992). Kinetics of HIV-1 long terminal repeat trans-activation—use of intragenic ribozyme to assess rate-limiting steps. J. Biol. Chem. 267, 17891–17899.

Johnson, D.B. (1989). Induction of ornithine decarboxylase by mitogen is affected by antioxidants: A free-radical mechanism in mitogenesis. Biochem. Soc. Trans. 17, 406–407.

Jurgensmeier, J.M., Schmitt, C.P., Viesel, E., Hofler, P., & Bauer, G. (1994). Transforming growth factor-β-treated fibroblasts eliminate transformed fibroblasts by induction of apoptosis. Cancer Res. 54, 393–398.

Kalebic, T., Kinter, A., Poli, G., &erson, M.E., Meister, A., & Fauci, A.S. (1991). Suppression of human immunodeficiency virus expression in chronically infected monocytic cells by glutathione, glutathione ester, and N-acetylcysteine. Proc. Natl. Acad. Sci. USA 88, 986–990.

Kane, D.J., Sarafian, T.A., Anton, R., Hahn, H., Gralla, E.B., Valentine, J.S., Ord, T., & Bredesen, D.E. (1993). Bcl-2 inhibition of neural death—Decreased generation of reactive oxygen species. Science 262, 1274–1277.

Kessler, M., & Mathews, M.B. (1992). Premature termination and processing of human immunodeficiency virus type 1-promoted transcripts. J. Virol. 66, 4488–4496.

Kim, C.M., Vogel, J., Jay, G., & Rhim, J.S. (1992). The HIV *tat* gene transforms human keratinocytes. Oncogene. 7, 1525–1529.

Kolson, D.L., Collman, R., Hrin, R., Balliet, J.W., Laughlin, M., Mcgann, K.A., Debouck, C., & Gonzalez-Scarano, F. (1994). Human immunodeficiency virus type 1 Tat activity in human neuronal cells: Uptake and trans-activation. J. Gen. Virol. 75, 1927–1934.

Kubota, S., el-Farrash, M.A., Maki, M., Harada, S., & Hatanaka, M. (1990). 2,3 Dimercapto-1-propanol inhibits HIV-1 tat activity, viral production, and infectivity *in vitro*. AIDS Res. Hum. Retroviruses 6, 919–927.

Laspia, M.F., Rice, A.P., & Mathews, M.B. (1989). HIV-1 Tat protein increases transcriptional initiation and stabilizes elongation. Cell 59, 283–292.

Laurence, J., & Astrin, S.M. (1991). Human immunodeficiency virus induction of malignant transformation in human B lymphocytes. Proc. Natl. Acad. Sci. USA 88, 7635–7639.

Leguern, M., Shioda, T., Levy, J.A., & Chengmayer, C. (1993). Single amino acid change in Tat determines the different rates of replication of 2 sequential HIV-1 isolates. Virology 195, 441–447.

Li, C.J., Friedman, D.J., Wang, C.L., Metelev, V., & Pardee, A.B. (1995). Induction of apoptosis in uninfected lymphocytes by HIV-1 Tat protein. Science 268, 429–431.

Liu, J.S., Perkins, N.D., Schmid, R.M., & Nabel, G.J. (1992). Specific NF-κB subunits act in concert with Tat to stimulate human immunodeficiency virus type-1 transcription. J. Virol. 66, 3883–3887.

Longo, F., Marchetti, M.A., Castagnoli, L., Battaglia, P.A., & Gigliani, F. (1995). A novel approach to protein–protein interaction: Complex formation between the P53 tumor suppressor and the HIV Tat proteins. Biochem. Biophys. Res. Commun. 206, 326–334.

Lu, Q.L., Hanby, A.M., Hajibagheri, M.A.N., Gschmeissner, S.E., Lu, P.J., Taylorpapadimitriou, J., Krajewski, S., Reed, J.C., & Wright, N.A. (1994). Bcl-2 protein localizes to the chromosomes of mitotic nuclei and is correlated with the cell cycle in cultured epithelial cell lines. J. Cell Sci. 107, 363–371.

Maeda, H., & Akaike, T. (1991). Oxygen free radicals as pathogenic molecules in viral diseases. Proc. Soc. Exp. Biol. Med. 198, 721–727.

Magnuson, D.S.K., Knudsen, B.E., Geiger, J.D., Brownstone, R.M., & Nath, A. (1995). Human immunodeficiency virus type 1 Tat activates non-N- methyl-D-aspartate excitatory amino acid receptors and causes neurotoxicity. Ann. Neurol. 37, 373–380.

Marciniak, R.A., Garcia-Blanco, M.A., & Sharp, P.A. (1990). Identification and characterization of a HeLa nuclear protein that specifically binds to the trans-activation-response (TAR) element of human immunodeficiency virus. Proc. Natl. Acad. Sci. USA 87, 3624–3628.

McCord, J.M., Boyle, J.A., Day, E.D.,Jr., Rizzolo, L.J., & Salin, M.L. (1977). A manganese-containing superoxide dismutase from human liver. In Superoxide and Superoxide Dismutase. Michelson, A.M., McCord, J.M., & Fridovich, I., Eds., pp. 129–138. Academic Press, London.

McCord, J.M. (1985). Oxygen-derived free radicals in post-ischemic tissue injury. N. Eng. J. Med. 312, 159–163.

Mcgrath, M.S., Shiramizu, B.T., & Herndier, B.G. (1995). Identification of a clonal form of HIV in early Kaposi's Sarcoma: Evidence for a novel model of oncogenesis, "sequential neoplasia." J. Acq. Immun. Defic. Synd. Hum. R. 8, 379–385.

Mihm, S., Ennen, J., Pessara, U., Kurth, R., & Droge, W. (1991). Inhibition of HIV-1 replication and NF-kapppa-B activity by cysteine and cysteine derivatives. AIDS 5, 497.

Milani, D., Zauli, G., Neri, L.M., Marchisio, M., Previati, M., & Capitani, S. (1993). Influence of the human immunodeficiency virus type-1 Tat protein on the proliferation and differentiation of PC12 rat pheochromocytoma cells. J. Gen. Virol. 74, 2587–2594.

Mosier, D.E., Gulizia, R.J., MacIsaac, P.D., Torbett, B.E., & Levy, J.A. (1993). Rapid loss of CD4+ T cells in human-PBL-SCID mice by noncytopathic HIV isolates. Science 260, 689–692.

Muesing, M.A., Smith, D.H., & Capon, D.J. (1987). Regulation of mRNA accumulation by a human immunodeficiency virus trans-activator protein. Cell 48, 691–701.

Muller, W.E., Okamoto, T., Reuter, P., Ugarkovic, D., & Schroder, H.C. (1990). Functional characterization of Tat protein from human immunodeficiency virus. Evidence that Tat links viral RNAs to nuclear matrix. J. Biol. Chem. 265, 3803–3808.

Nakamura, Y., Gindhart, T., Winterstein, D., Tomita, I., Seed, J., & Colburn, N. (1988). Early superoxide dismutase-sensitive event promotes neoplastic transformation in mouse epidermal JB6 cells. Carcinogenesis 9, 203–207.

Nakashima, J.M., Levin, J.R., Hyde, D.M., & Giri, S.N. (1991). Repeated exposures to enzyme-generated oxidants cause alveolitis, epithelial hyperplasia, and fibrosis in hamsters. Am. J. Pathol. 139, 1485–1499.

Nishioka, W.K., & Welsh, R.M. (1994). Susceptibility to cytotoxic T lymphocyte-induced apoptosis is a function of the proliferative status of the target. J. Exp. Med., 179, 769–774.

Nose, K., Shibanuma, M., Kikuchi, K., Kageyama, H., Sakiyama, S., & Kuroki, T. (1991). Transcriptional activation of early-response genes by hydrogen peroxide in a mouse osteoblastic cell line. Eur. J. Biochem., 201, 99–106.

Ojeda, F., Guarda, M.I., Maldonado, C., Folch, H., & Diehl, H. (1992). Role of protein kinase-C in thymocyte apoptosis induced by irradiation. Int. J. Radiat. Biol., 61, 663–667.

Omar, R.A., Yano, S., & Kikkawa, Y. (1987). Antioxidant enzymes and survival of normal and simian virus 40-transformed mouse embryo cells after hyperthermia. Cancer Res., 47, 3473–3476.

Pantaleo, G., Menzo, S., Vaccarezza, M., Graziosi, C., Cohen, O.J., Demarest, J.F., Montefiori, D., Orenstein, J.M., Fox, C., Schrager, L.K., & et al, (1995). Studies in subjects with long-term nonprogressive human immunodeficiency virus infection [see comments]. N. Engl. J. Med. 332, 209–216.

Peterhans, E., Jungi, T.W., & Stocker, R. (1988). Autotoxicity and reactive oxygen in viral disease. In: Oxy-Radicals in Molecular Biology and Pathology. (Cerutti, P.A., Fridovich, I., & McCord, J.M., Eds.), pp. 543–562. Alan R. Liss, New York.

Purvis, S.F., Georges, D.L., Williams, T.M., & Lederman, M.M. (1992). Suppression of interleukin-2 and interleukin-2 receptor expression in Jurkat cells stably expressing the human immunodeficiency virus Tat protein. Cell Immunol. 144, 32–42.

Purvis, S.F., Jacobberger, J.W., Sramkoski, R.M., Patki, A.H., & Lederman, M.M. (1995). HIV type 1 Tat protein induces apoptosis and death in Jurkat cells. AIDS Res. Hum. Retroviruses, 11, 443–450.

Radman, M. (1980). Is there SOS induction in mammalian cells? Photochem. Photobiol. 32, 823–830.

Ralph, W.M.,Jr., Cabatingan, M.S., & Schaffer, P.A. (1994). Induction of herpes simplex virus type 1 immediate–early gene expression by a cellular activity expressed in Vero and NB41A3 cells after growth arrest-release. J. Virol. 68, 871–6882.

Ramakrishnan, N., & Catravas, G.N. (1992). N-(2-mercaptoethyl)-1,3-propanediamine (WR-1065) protects thymocytes from programed cell death. J. Immunol. 148, 1817–1821.

Rao, G.N., & Berk, B.C. (1992). Active oxygen species stimulate vascular smooth muscle cell growth and proto-oncogene expression. Circ. Res. 70, 593-599.

Rautonen, J., Rautonen, N., Martin, N.L., & Wara, D.W. (1994). HIV type 1 Tat protein induces immunoglobulin and interleukin 6 synthesis by uninfected peripheral blood mononuclear cells. AIDS Res. Hum. Retroviruses 10, 781–785.

Richter, C. (1993). Pro-oxidants and mitochondrial Ca^{2+}—Their relationship to apoptosis and oncogenesis. FEBS Lett. 325, 104–107.

Roederer, M., Staal, F.J.T., Raju, P.A., Ela, S.W., & Herzenberg, L.A. (1990). Cytokine-stimulated human immunodeficiency virus replication is inhibited by N-acetyl-L-cysteine. Proc. Natl. Acad. Sci. USA 87, 4884–4888.

Rubin, L.L., Philpott, K.L., & Brooks, S.F. (1995). The cell cycle and cell death. Curr. Biol. 3, 391–394.

Sadaie, M.R., Mukhopadhyaya, R., Benaissa, Z.N., Pavlakis, G.N., & Wong-Staal, F. (1990). Conservative mutations in the putative metal-binding region of human immunodeficiency virus *tat* disrupt virus replication. AIDS Res. Hum. Retroviruses, 6, 1257–1263.

Sandstrom, P.A., Mannie, M.D., & Buttke, T.M. (1994a). Inhibition of activation-induced death in T cell hybridomas by thiol antioxidants—Oxidative stress as a mediator of apoptosis. J. Leukocyte. Biol., 55, 221–226.

Sandstrom, P.A., Tebbey, P.W., Vancleave, S., & Buttke, T.M. (1994b). Lipid hydroperoxides induce apoptosis in T-cells displaying a HIV-associated glutathione peroxidase deficiency. J. Biol. Chem. 269, 798–801.

Sastry, K.J., Reddy, R.H.R., Pandita, R., Totpal, K., & Aggarwal, B.B. (1990). HIV-1 *tat* gene induces tumor necrosis factor-beta(lymphotoxin) in a human B-lymphoblastoid cell line. J. Biol. Chem., 265, 20091–20093.

Scheppler, J.A., Nicholson, J.K.A., Swan, D.C., Ahmed-Ansari, A., & McDougal, J.S. (1989). Down-modulation of MHC-I in a $CD4^+$ T cell line, CEM-E5, after HIV-1 infection. J. Immunol. 143, 2858–2866.

Schreck, R., Rieber, P., & Baeuerle, P.A. (1991). Reactive oxygen intermediates as apparently widely used messengers in the activation of the NF-kappa B transcription factor and HIV-1. EMBO J., 10, 2247–2258.

Schreck, R., Meier, B., Mannel, D.N., Droge, W., & Baeuerle, P.A. (1992). Dithiocarbamates as potent inhibitors of nuclear factor kappa B activation in intact cells. J. Exp. Med., 175, 1181–1194.

Selby, M.J., & Peterlin, B.M. (1990). Trans-activation by HIV-1 *tat* via a heterologous RNA-binding protein. Cell 62, 769–776.

Sentman, C.L., Shutter, J.R., Hockenbery, D., Kanagawa, O., & Korsmeyer, S.J. (1991). Bcl-2 inhibits multiple forms of apoptosis but not negative selection in thymocytes. Cell 67, 879–888.

Shi, Y., Glynn, J.M., Guilbert, L.J., Cotter, T.G., Bissonnette, R.P., & Green, D.R. (1992). Role for c-myc in activation-induced apoptotic cell death in T cell hybridomas. Science 257, 212–214.

Southgate, C.D., & Green, M.R. (1991). The HIV-1 Tat protein activates transcription from an upstream DNA-binding site: implications for Tat function. Genes Dev. 5, 2496–2507.

Staal, F.J.T., Ela, S.W., Roederer, M., &erson, M.T., & Herzenberg, L.A. (1992). Glutathione deficiency and human immunodeficiency virus infection. Lancet 339, 909–912.

Stanley, S.K., Folks, T.M., & Fauci, A.S. (1989). Induction of expression of human immunodeficiency virus in a chronically infected promonocytic cell line by ultraviolet irradiation. AIDS Res. Hum. Retroviruses 5, 375–384.

Stein, B., Rahmsdorf, H.J., Steffen, A., Litfin, M., & Herrlich, P. (1989). UV-induced DNA damage is an intermediate step in UV-induced expression of human immunodeficiency virus type 1, collagenase, *c-fos*, and metallothionein. Mol. Cell Biol. 9, 5169–5181.

Steinman, H.M. (1995). The bcl-2 oncoprotein functions as a pro-oxidant. J. Biol. Chem. 270, 3487–3490.

Steller, H. (1995). Mechanisms and genes of cellular suicide. Science 267, 1445–1449.

Storz, G., Tartaglia, L.A., & Ames, B.N. (1990). Transcriptional regulator of oxidative stress-inducible genes: Direct activation by oxidation. Science 248, 189–194.

Taylor, J.P., Cupp, C., Diaz, A., Chowdhury, M., Khalili, K., Jimenez, S.A., & Amini, S. (1992a). Activation of expression of genes coding for extracellular matrix proteins in tat-producing glioblastoma cells. Proc. Natl. Acad. Sci. USA 89, 9617–9621.

Taylor, J.P., Pomerantz, R., Bagasra, O., Chowdhury, M., Rappaport, J., Khalili, K., & Amini, S. (1992b). TAR-independent transactivation by Tat in cells derived from the CNS—A novel mechanism of HIV-1 gene regulation. EMBO J. 11, 3395–3403.

Thomas, C.E., Morehouse, L.A., & Aust, S.D. (1985). Ferritin and superoxide-dependent lipid peroxidation. J. Biol. Chem. 260, 3275–3280.

Tiley, L.S., Madore, S.J., Malim, M.H., & Cullen, B.R. (1992). The VP16 transcription activation domain is functional when targeted to a promoter-proximal RNA sequence. Gene Dev. 6, 2077–2087.

Toledano, M.B., & Leonard, W.J. (1991). Modulation of transcription factor NF-kappa-B binding activity by oxidation–reduction in vitro. Proc. Natl. Acad. Sci. USA 88, 4328–4332.

Ursini, F., Maiorino, M., & Gregolin, C. (1985). The selenoenzyme phospholipid hydroperoxide glutathione peroxidase. Biochim. Biophys. Acta 839, 62–70.

Vaishnav, Y.N., & Wong-Staal, F. (1991). The biochemistry of AIDS. Ann. Rev. Biochem. 60, 577–630.

Virelizier, J-L. (1990). Cellular activation and human immunodeficiency virus infection. Curr. Opin. Immunol. 2, 409.

Viscidi, R.P., Mayur, K., Lederman, H.M., & Frankel, A.D. (1989). Inhibition of antigen-induced lymphocyte proliferation by Tat protein from HIV-1. Science 246, 1606–1608.

Vogel, J., Hinrichs, S.H., Reynolds, R.K., Luciw, P.A., & Jay, G. (1988). The HIV *tat* gene induces dermal lesions resembling Kaposi's sarcoma in transgenic mice. Nature 335, 606–611.

Walker, P.R., Kwast-Welfeld, J., Gourdeau, H., Leblanc, J., Neugebauer, W., & Sikorska, M. (1993). Relationship between apoptosis and the cell cycle in lymphocytes: roles of protein kinase C, tyrosine phosphorylation, and AP1. Exp. Cell Res. 207, 142–151.

Wang, Y.J., & Watson, R.R. (1994). Potential therapeutics of vitamin E (Tocopherol) in AIDS and HIV. Drugs. 48, 327–338.

Weeks, B.S., Desai, K., Loewenstein, P.M., Klotman, M.E., Klotman, P.E., Green, M., & Kleinman, H.K. (1993). Identification of a novel cell attachment domain in the HIV-1 Tat protein and its 90-kDa cell surface binding protein. J. Biol. Chem. 268, 5279–5284.

Westendorp, M.O., Shatrov, V.A., Schulzeosthoff, K., Frank, R., Kraft, M., Los, M., Krammer, P.H., Droge, W., & Lehmann, V. (1995). HIV-1 Tat potentiates TNF-induced NF-kappa B activation and cytotoxicity by altering the cellular redox state. EMBO J. 14, 546–554.

Wong, G.H.W., McHugh, T., Weber, R., & Goeddel, D.V. (1991). Tumor necrosis factor α selectively sensitizes human immunodeficiency virus-infected cells to heat and radiation. Proc. Natl. Acad. Sci. USA 88, 4372–4376.

Wu, F., Garcia, J., Sigmin, D., & Gaynor, R. (1991). Tat regulates binding of the human immunodeficiency virus trans-activating region RNA loop-binding protein TRP-185. Genes. Dev. 5, 2128–2140.

Yamada, T., Yamaoka, S., Goto, T., Nakai, M., Tsujimoto, Y., & Hatanaka, M. (1994). The human T-cell leukemia virus type I Tax protein induces apoptosis, which is blocked by the Bcl-2 protein. J. Virol. 68, 3374–3379.

Yamanaka, N., & Deamer, D. (1974). Superoxide dismutase activity in WI-38 cell cultures: Effects of age, trypsinization and SV–40 transformation. Physiol. Chem. Phys. 6, 95–106.

Yasunaga, T., Kato, H., Ohgaki, K., Inamoto, T., & Hikasa, Y. (1982). Effect of vitamin E as an immunopotentiation agent for mice at optimal dosage and its toxicity at high dosage. J. Nutr. 112, 1075–1084.

Zauli, G., Gibellini, D., Milani, D., Mazzoni, M., Borgatti, P., Laplaca, M., & Capitani, S. (1993). Human immunodeficiency virus type-1 Tat protein protects lymphoid, epithelial, and neuronal cell lines from death by apoptosis. Cancer Res. 53, 4481–4485.

NEUTROPHILS AND ISCHEMIC/REPERFUSION INJURY

Norman R. Harris and D. Neil Granger

Advances in Molecular and Cellular Biology
Volume 25, pages 95-111.

ISBN: 0-7623-0379-4

I. ISCHEMIC INJURY VERSUS REPERFUSION INJURY

Tissue necrosis is the ultimate consequence of uninterrupted ischemia since low blood flow results in cellular oxygen deprivation. Cell viability is dependent on mechanisms utilizing the energy of ATP, a molecule that requires oxygen for its production. The length of time necessary for irreversible injury due to a lack of oxygen depends on the tissue. Severe damage results from just minutes of ischemia in the brain, while skeletal muscle can withstand hours of such hypoxia. Therefore, the clinical aim in relieving ischemic trauma is promptly restoring blood flow. However, even though reperfusion can avert immediate irreversible tissue damage, a considerable portion, if not the majority of the eventual injury can occur after blood flow has been restored.

Such reperfusion injury was demonstrated in a study in which the small intestine of the cat was rendered ischemic for either 4 hours with no reperfusion or for 3 hours with 1 hour of reperfusion (Parks and Granger, 1986). The results indicated that mucosal inflammation and necrosis was more severe when the ischemia was interrupted by reperfusion. A similar study was performed by placing a tourniquet around both hind limbs in a model of I/R in the rat (Seekamp et al., 1993). Eight hours of ischemia produced no increase in permeability of the skeletal muscle microvasculature (assessed by leakage of albumin). However, when 4 hours of ischemia was followed by 4 hours of reperfusion, vascular permeability increased four-fold.

More evidence of the relative importance of reperfusion injury comes from several studies where drug administration at the time of reperfusion was found to be as effective as that given at the onset of ischemia. These will be discussed later in the chapter. One such therapy that is receiving considerable attention is the administration of drugs that scavenge the excess oxidants that are produced at the time of reperfusion.

II. DELETERIOUS EFFECT OF REOXYGENATING ISCHEMIC TISSUE

Reoxygenation of ischemic tissue can result in the formation of oxygen free radicals, molecules with an unpaired valence electron and thus with significant chemical reactivity. A proposed mechanism (Granger, 1988) by which oxidants are formed during I/R begins with the breakdown of cellular ATP to hypoxanthine (HX). This process is accompanied by a conversion of xanthine dehydrogenase (XDH) to xanthine oxidase (XO). XO catalyzes the reaction between HX and the O_2 supplied by reperfusion to form the superoxide radical, O_2^-, and hydrogen peroxide, H_2O_2. While not a free-radical by definition, H_2O_2 has been called an "honorary" free-radical due to its potent oxidizing ability. These oxidants may directly cause some degree of microvascular damage, but evidence is accumulating that this is only an intermediate stage in the injury mechanism. The process continues as the hydroxyl radical (·OH) is be formed through the iron-catalyzed reaction between O_2^- and H_2O_2. The ·OH radical in turn can cause lipid peroxidation resulting in the stimulation of neutrophils, which produce oxidants and proteases capable of inflicting direct injury to the microvascular endothelium. This mechanism is summarized in Figure 1.

Even though the body possesses scavengers and antioxidants that are able to neutralize how to moderate levels of oxidant production, these defense mechanisms appear to be unable to control the sudden excess of radicals produced in postischemic tissues. To make matters worse, the activity of endogenous antioxidant enzymes such as superoxide dismutase (SOD) and catalase actually decreases during ischemia (Grisham et al., 1986). Therefore, considerable research has been done to determine whether or not administration of additional antioxidants during I/R can prevent injury.

Each of the antioxidants in Figure 1 have been shown to provide at least partial protection from the deleterious effects of I/R. Allopurinol and oxypurinol prevent superoxide formation through an inhibitory effect on XO, and to a lesser extent by scavenging free radicals. SOD and catalase each have been found to reduce I/R injury: SOD catalyzes the reduction of O_2^- to H_2O_2, while catalase breaks H_2O_2 down to H_2O. Often SOD and catalase are administered simultaneously to eliminate both oxidant species. While these oxygen free radicals have been shown to be involved in I/R-induced tissue damage, their production cannot entirely explain the overall injury process; evidence is accumulating that neutrophils also make a significant contribution.

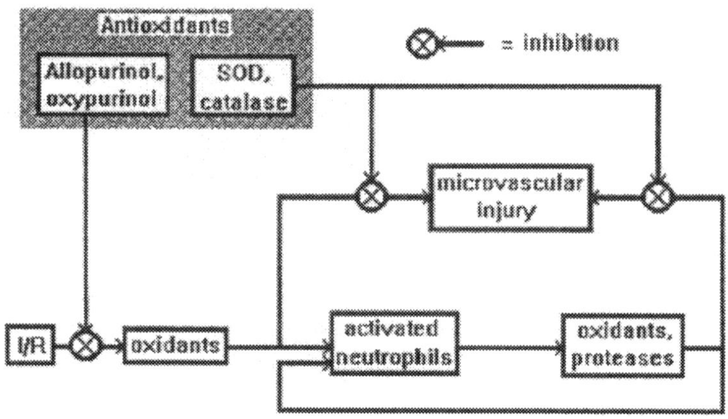

Figure 1. Mechanism where I/R-induced injury arises from oxidants produced by I/R itself and also from subsequent neutrophil activation. Oxidant formation is reduced by the XO inhibitors allopurinol and oxypurinol; SOD and catalase scavenge oxidants formed by I/R.

III. ARE NEUTROPHILS INVOLVED IN I/R INJURY?

A consistent finding in the investigation of I/R injury is the tissue accumulation of polymorphonuclear leukocytes (neutrophils). However, in several models of I/R, the question remains whether the infiltration is a cause or merely an effect of the injury itself. One approach that has been used to determine the role of the neutrophil is to eliminate the majority of these circulating cells from the experimental animal prior to the I/R protocol. For example, in a study of the small intestine of the cat (Hernandez et al., 1987), one hour of ischemia followed by two hours of reperfusion increased the permeability parameter, 1-σ (where σ is the osmotic reflection coefficient), from a baseline value of 0.08 to a significantly higher value of 0.32. This increase was almost entirely abolished in cats that had been pretreated with antineutrophil serum (ANS). The same parameter of permeability was monitored in a similar study of canine skeletal muscle (Carden et al., 1990). Four hours of ischemia followed by 30 minutes of reperfusion resulted in more than a two-fold increase in 1-σ, while dogs pretreated with ANS displayed no increase in 1-σ in response to I/R.

Similar results are obtained in the rat kidney (Hellberg et al., 1990), where 45 minutes of ischemia followed by 1 hour of reperfusion resulted in a two-fold increase in a protein permeability index, while no increase was observed in animals pretreated with ANS. In another model, neutropenia in the rat significantly reduced edematous brain water content in response to a stroke model in which 1 hour of cerebral ischemia was followed by a 24 hour reperfusion period (Shiga et al., 1991).

Models of coronary I/R also supply evidence of a role for neutrophils. The extent of canine myocardial infarcts were compared between groups with or without pretreatment with ANS (Romson et al., 1983). Infarcts due to 90 minutes of ischemia then 6 hours of reperfusion were 43% smaller in the ANS group. Infarct size was 67% smaller than the control I/R group in a similar canine model (1 hour of ischemia followed by 5 hours of reperfusion) when leukopenia was induced by hydroxyurea (Mullane et al., 1984). In this case the protection was not specifically due to neutrophil inhibition since hydroxyurea doesn't possess the same specificity as ANS and thus indiscriminately eliminates several types of white cells including neutrophils.

Another organ protected by ANS treatment in a model of I/R is the liver (Jaeschke et al., 1990). Forty-five minutes of complete hepatic ischemia in the rat followed by 24 hours of reperfusion resulted in 80% necrosis. However, when rats were rendered neutropenic with ANS, the value decreased significantly to 28%.

The injurious role of neutrophils has also been investigated in the "whole body" model of I/R, hemorrhagic shock and resuscitation. In a rat study of shock (Smith et al., 1987), blood was withdrawn so that arterial pressure was lowered to 27 mm Hg for 30 minutes. At that time the shed blood was reinfused, and an hour later the amount of ^{51}Cr-labeled red blood cells that had leaked into the gastric lumen was quantitated. Red blood cell leakage was decreased by a factor of five by pretreatment of the rats with ANS.

Two even more severe models of hemorrhagic shock give additional evidence that neutrophils are involved in I/R injury (Barroso-Aranda et al., 1988). In the first, arterial pressure in rats was lowered to 40 mm Hg for three hours. Only 36% survived without ANS treatment compared to 100% with ANS. When the survival rate was decreased to 0% by an even more severe shock treatment (30 mm Hg for 7 hours), the survival rate remained at 100% in the animals given ANS.

While studies of neutropenia give a compelling confirmation of a significant role for neutrophils in I/R, this is not the only available evidence. In order for neutrophils to become activated to release their injury-causing oxidants and proteases, it appears that they must first become adherent to the endothelial cells that line the blood vessel walls. As will be discussed later, antibodies which interfere with this adhesion process are able to significantly diminish injury associated with I/R. In order to understand how these antibodies afford such protection, a discussion of leukocyte–endothelial cell adhesion and its relevance to I/R is warranted.

IV. I/R-INDUCED NEUTROPHIL RECRUITMENT

A. Extent of Neutrophil Infiltration

It is now well established that leukocytes accumulate in areas affected by I/R. This was demonstrated in a few of the studies already mentioned in the previous section of this chapter. For example, in the rat kidney, 45 minutes of ischemia followed by one hour of reperfusion caused an approximate 10-fold increase in tissue neutrophils compared to control values (Hellberg, 1990). The same duration of ischemia and reperfusion in the rat liver produced a five-fold increase in neutrophil infiltration (Jaeschke, 1990). This phenomena is widespread throughout the organs of the body with similar increases in neutrophil accumulation in a relatively short period of reperfusion: 24 hours of ischemia followed by two hours of reperfusion caused an approximate three-fold increase in rabbit lungs (Horgan, 1991); and similar neutrophil infiltration occurs in the intestine (Grisham et al., 1986), skeletal muscle (Rubin et al., 1990; Korthuis et al., 1991), and in the heart (Tanaka et al., 1993).

The magnitude of I/R-induced neutrophil infiltration increases with the duration of reperfusion. This is shown in two separate studies of I/R-induced neutrophil accumulation in canine skeletal muscle. Four hours of ischemia followed by 30 minutes of reperfusion (Korthuis, 1991) induced an approximate 20-fold increase in tissue myeloperoxidase (MPO) activity (a measure of neutrophil infiltration) while a similar period of ischemia (five hours) followed by a much longer reperfusion period of 48 hours resulted in a 97-fold increase in MPO (Rubin, 1990). This trend was also observed in the cat small intestine (Grisham, 1986) and in the rat liver (Jaeschke, 1990). In cat intestine, tissue MPO activity continued to increase following 3 hours of ischemia: at the onset of reperfusion, MPO activity increased five-fold, then 10-fold at 15 minutes of reperfusion, and 17-fold at 1 hour of reperfusion. In rat liver, neutrophil infiltration increased in a similar manner following 45 minutes of ischemia: a five-fold increase at one hour of reperfusion, a 17-fold increase at 6 hours, and an 80-fold increase at 24 hours.

The neutrophil accumulation is not necessarily localized to the organ subjected to the ischemia and reperfusion. For example, in a study of I/R in the rat intestine, *lung* MPO activity increased four-fold in response to 2 hours of ischemia and seven-fold following 15 minutes of reperfusion (Schmeling, 1989). Similarly, 4 hours of ischemia in the rat hind limb doubled MPO activity in the lung, while 4 hours of reperfusion caused a five-fold pulmonary MPO increase (Seekamp, 1993).

From these studies and many others, it is obvious that the I/R-induced neutrophil sequestration can be quite profound. Investigators have reasoned from a therapeutic standpoint that if neutrophil infiltration contributes to the pathophysiology of I/R injury, then interference with this mechanism could be clinically desirable. Much research has been done in the past decade in an attempt to identify the molecular and hemodynamic factors that regulate the adherence and emigration of neutrophils in the microcirculation and to determine whether or not attenuating these adhesion-dependent events will reduce I/R injury.

B. Mechanism of Neutrophil Infiltration

Figure 2 illustrates the five main steps in the process whereby a neutrophil makes its way from the bloodstream into the tissue: (1) margination from the middle of the blood stream toward the vessel wall, (2) rolling along the vessel wall, (3) firm adherence to the endothelial cells that line the vessel wall, (4) extravasation through endothelial intercellular junctions, and (5) emigration from the vessel into the tissue.

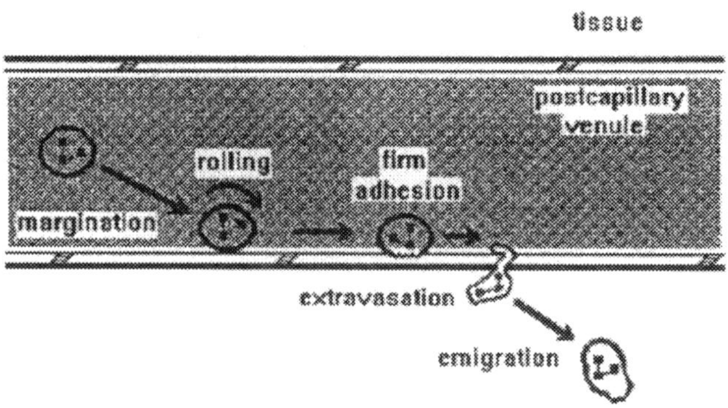

Figure 2. Steps of tissue recruitment of leukocytes: Margination to the vascular wall, rolling along the endothelium, firm adhesion to endothelial cells, extravasation through intercellular gaps, and emigration into the interstitium.

While the majority of the knowledge of the molecular mechanisms involved in leukocyte adhesion has been determined in the past decade, its existence has been recognized for well over 100 years. Knowledge of leukocyte margination along vessel walls prior to 1938 was summarized in a doctoral dissertation by Gert Vejlens (Vejlens, 1938). At that time, it was known that leukocyte margination is an event primarily found in *post*capillary venules, and that hemodynamic factors influence their position in the bloodstream. He describes work by R. Fahraeus who explained that "In the competition for place in the axial stream the white blood corpuscles are pushed out towards the periphery of the blood vessel by the large aggregates of red blood corpuscles and become marginal."

Hemodynamic principles alone predict that the displaced leukocytes would travel down the margins of the vessel walls at a greater speed than which they are actually observed. It is now known that molecules found on both the leukocytes and vascular endothelial cells interact with each other, inducing an adhesive effect that results in leukocyte rolling and firm adhesion. Additionally, such adhesion molecules on the endothelial cells are much more prominent on the postcapillary venules than in other sections of the microvasculature, accounting for the observation that adhesion occurs primarily downstream of the capillaries.

An endothelial adhesion molecule that accounts for a large portion of the low-affinity binding manifested as leukocyte rolling is P-selectin. P-selectin has been shown to bind to the sialyl-Lewis X oligosaccharide normally found on the surface of leukocytes. Two other selectins with adhesive capabilities are E-selectin on the endothelial cell and L-selectin on the leukocyte. The close proximity of the two cell types induced by rolling allows interaction between the endothelial adhesion molecule ICAM-1 and the leukocyte complex CD11/CD18. The CD11/CD18 heterodimer has three different α-subunits identified as CD11a (found on all leukocytes) and CD11b and CD11c (found primarily on neutrophils and monocytes) each bound to the common β-subunit CD18. The glycoprotein ICAM-1 is constitutively expressed on the endothelial surface and its expression is upregulated during I/R. The interaction between CD11/CD18 and ICAM-1 is relatively strong, often resulting in "firm" adhesion, which lasts from a few seconds to perhaps 30 minutes. Once neutrophils become firmly adherent to the endothelium, they may migrate to an intercellular cleft, where they are able to change shape dramatically to "squeeze" through the gap and move away from the vessel wall into the interstitium.

C. Leukocyte Adherence and Emigration Initiated by I/R

The transparency of certain thin tissues allows for microscopic observation and quantitation of leukocyte–endothelial cell adhesion in the microcirculation. Examples of such tissues that have been used for this purpose include the cheek pouch of the hamster and the mesentery of several different species. Our laboratory has performed numerous studies in which the cat and rat mesentery have been exteriorized and observed by *in vivo* microscopy, and images from this procedure have been published (Kubes et al., 1990; Suzuki et al., 1991b). In one of these studies, the adherence and emigration of leukocytes in response to hemorrhagic shock was observed in cat mesenteric venules (Perry and Granger, 1992). One hour of ischemia produced an approximate three-fold increase in both the number of adherent and emigrated leukocytes. One hour of reperfusion exacerbated the inflammatory response with both the number of adherent and emigrated leukocytes approximately five-fold higher than the values obtained during baseline conditions. This study also demonstrated that the leukocyte adherence and emigration responses to 1 hour each of local ischemia (induced by partial occlusion of the superior mesenteric artery) and reperfusion were very similar to results obtained with I/R elicited by hemorrhage-retransfusion, in other words, in both instances adherence and extravasation increased approximately five-fold.

The results by Perry and Granger (1992) are similar to previous results in the cat mesentery where one hour of local intestinal ischemia followed by one hour of reperfusion caused a six-fold increase in adherence and an eight-fold increase in extravasation (Granger et al., 1989). Also, the increases were six-fold and seven-fold in adherence and emigration in another study using the same protocol in the cat mesentery (Oliver et al., 1991). However, these leukocyte-endothelial cell interactions are not unique to the mesentery: dramatic increases in I/R-induced leukocyte adherence have been observed in similar microscopic observations of the hamster cheek pouch (Erlansson et al., 1991; Bertuglia et al., 1993) and in a dorsal skin fold chamber in the hamster (Lehr et al., 1991). These models have also been used to provide clear evidence for an inhibitory effect of antioxidants on I/R-induced leukocyte adhesion.

D. Oxidant Contribution to I/R-Induced Adherence and Emigration

Using the model of local intestinal ischemia and reperfusion in the cat mesentery, the role of oxidants in I/R-induced adherence and emigration has been investigated using CuZn-SOD and catalase as well as the XO inhibitors allopurinol and oxypurinol. In each of the following three studies, the mesentery was prepared for *in vivo* video microscopy prior to 1 hour of ischemia (flow reduced to 20% of baseline values by occlusion of the superior mesenteric artery) and one hour of reperfusion.

In the first study (Granger et al., 1989), cats were treated with either CuZn-SOD, allopurinol, or with no drug as a control group. While the I/R protocol increased leukocyte-endothelial cell adherence by a factor of six in the control group, allopurinol and SOD inhibited this response by 60% and 40%, respectively. Similar results were found with the emigration response: I/R induced an eight-fold increase that was inhibited by 40% and 60% with allopurinol and SOD, respectively.

In the other two studies, treatment was withheld until after 1 hour each of ischemia and reperfusion. In the first of these two studies (Suzuki et al., 1989), SOD was found to inhibit the I/R-induced adherence, reducing the value to 60% of that observed after the hour-long reperfusion period. This effect of post-reperfusion administration of Mn-SOD on leukocyte adherence was reproduced in a subsequent study (Suzuki et al., 1991b) where the effects of catalase and oxypurinol were also monitored. Again, one hour of both ischemia and reperfusion was followed by administration of one of the three drugs. The number of adherent leukocytes was reduced to 73% (catalase), 82% (SOD), and 73% (oxypurinol) of the adherence observed following 1 hour of reperfusion.

The results of these three studies indicate that oxidants contribute to I/R-induced adherence and emigration. An additional line of evidence that supports this view is provided by reports demonstrating that both superoxide and H_2O_2 are capable of inducing leukocyte adhesion in postcapillary venules. In one study (Del Maestro et al., 1982) superoxide and H_2O_2 were produced by superfusing the hamster cheek pouch with hypoxanthine (HX) and xanthine oxidase (XO). A significant increase in the number of adherent leukocytes in postcapillary venules was observed and this increase was attenuated by SOD. Also, since administration of catalase was unable to also reduce adhesion, superoxide rather than H_2O_2 was implicated as the direct cause of adhesion. However, H_2O_2 was demonstrated to induce adhesion in another study (Suzuki et al., 1991a) in postcapillary venules of the cat mesentery: Superfusion of H_2O_2 caused a seven-fold increase in the number of adherent leukocytes. Also noted in this study was the ability of an antibody against the CD18 molecule to significantly reduce (by 85%) the H_2O_2-induced adherence. As described in the following section, antibodies against CD18, as well as other adhesion molecules, are also capable of preventing the adherence and emigration elicited by I/R.

E. Role of Adhesion Molecules

Antibodies have been developed that bind to and functionally inactivate adhesion molecules on both endothelial cells and neutrophils. The most successful attempts at interfering with adhesion have been with antibodies against the common β subunit (CD18) of the CD11/CD18 complex on leukocytes. One such CD18-specific monoclonal antibody (MAb), IB_4, has been extensively studied in a feline model of I/R-induced leukocyte adherence in postcapillary venules.

In one study (Suzuki et al., 1989), flow through the mesenteric circulation was reduced by occlusion of the superior mesenteric artery. One hour of ischemia was followed by one hour of reperfusion, at which time MAb IB_4 was administered. Ten minutes thereafter, the number of adherent leukocytes was reduced by 76%, which is approximately twice as effective as SOD, when the enzyme was administered using their same protocol (see previous section). Although this degree of attenuation of leukocyte adherence by MAb IB_4 is substantial, a more profound effect is noted when IB_4 is administered prior to ischemia (Oliver et al., 1991). In cat mesenteric venules, treatment with MAb IB_4 prior to one hour of ischemia and one hour of reperfusion completely eliminated the six- to seven-fold increases in leukocyte adherence and emigration normally observed in untreated preparations exposed to I/R.

A recent study (Kurose et al., 1994) in the rat mesentery investigated the antiadhesive properties of MAbs directed against CD18, CD11, ICAM-1, P-selectin, E-selectin, and L-selectin. The superior mesenteric artery was occluded for 20 minutes prior to a 30 minute reperfusion period, and levels of leukocyte adherence, emigration, and leakage of fluorescently-labeled albumin increased six, eight, and 10-fold, respectively, when no MAb was administered. The two most effective MAbs in inhibiting the increases in these three parameters were against CD18 and CD11b, attenuating the responses by 60 to 80%. The next most effective was the ICAM-1 MAb, which inhibited the responses by 40–60%. While the L-selectin MAb also demonstrated inhibition 10 minutes into reperfusion, none of three selectin MAbs were as effective as the CD11b, CD18, and ICAM-1 MAbs in reducing adherence, emigration, and protein leakage 30 minutes after reperfusion. However, P-selectin provided the greatest attenuation (80%) of the six MAbs when the I/R-induced platelet-leukocyte thrombi formation was quantitated. The CD11b, CD18, and ICAM-1 MAbs also reduced thrombus formation (by approximately 60%), while E-selectin and L-selectin provided no such inhibition.

Overall, the aforementioned studies demonstrate that the recruitment of leukocytes into postischemic tissues is modulated by adhesion molecules expressed on the surface of activated endothelial cells and leukocytes. Immunoneutralization of these adhesion receptors reduces I/R-induced leukocyte accumulation by interfering with the rolling, firm adhesion, and emigration of these cells in postcapillary venules. At the tissue level these alterations are manifested as a blunted increment in myeloperoxidase activity (Kurtel et al., 1992).

V. ROLE OF LEUKOCYTES IN I/R-INDUCED INJURY

A. Protection through Adhesion-Blocking Antibodies

The following studies describe the effects of different antibodies to adhesion glycoproteins on injury induced by either local or "whole-body" I/R. The CD11/CD18-ICAM-1 interaction that is thought to be responsible for much of the firm adhesion between leukocytes and endothelial cells has received the most attention from the perspective of preventive therapy.

Antibodies directed against the leukocyte adhesion glycoprotein CD11/CD18 have proven to be very effective in attenuating I/R injury. Two studies of I/R in cat intestine support this view. In the first study (Hernandez et al., 1987), occlusion of the superior mesenteric artery for one hour followed by two hours of reperfusion resulted in a four-fold increase in microvascular protein permeability ($1\text{-}\sigma$): during baseline conditions $1\text{-}\sigma = 0.08$ and after reperfusion $1\text{-}\sigma = 0.32$. Treatment of a separate group of cats with a CD18-specific MAb (60.3) prior to the I/R protocol attenuated the increase in $1\text{-}\sigma$: a value of 0.12 was obtained. These results were substantiated by a different group of investigators (Schoenberg et al., 1991) who also used a model of intestinal I/R in the cat. Two hours of ischemia followed by one hour of reperfusion induced a large increase in tissue MPO levels as well as mucosal lesions. However, each effect was significantly attenuated when MAb IB_4 was administered immediately before the reperfusion period.

The protective effects of CD18-specific MAbs are not limited to the intestine. Other models of local I/R produce the same results. For example, CD18-specific antibodies prevent myocardial I/R injury in the ferret (Gomoll et al., 1991), skeletal muscle I/R in the dog (Carden et al., 1990), and I/R injury in the rabbit ear (Vedder et al., 1990). In this latter study, an incision was made through the ear to allow occlusion of its main artery. Complete ischemia lasted 10 hours before reperfusion was allowed. Tissue injury was assessed by the degree of edematous weight gain in the ear. MAb 60.3 given either immediately before ischemia or immediately before reperfusion greatly attenuated the fluid accumulation observed in a control I/R group of rabbits.

Monoclonal antibodies directed against CD11/CD18 have been tested in several models of myocardial ischemia. In a study in the monkey (Winquist et al., 1990), 90 minutes of ischemia followed by four hours of reperfusion caused significant (27%) tissue necrosis with three of 15 animals dying from ventricular fibrillation. However, pretreatment with MAb R15.7 resulted in only 9% necrosis, with no mortality in 10 animals. In another study, myocardial ischemia in cats (90 minutes ischemia, 4.5 hours reperfusion) induced 31% necrosis and a significant elevation in tissue MPO levels. MAb R15.7 treatment once again reduced necrosis (to 2%) and also attenuated the increased MPO levels.

The CD18 MAbs also provide dramatic protection in models of hemorrhagic shock and subsequent resuscitation. In a study of shock in rabbits (Vedder, et al., 1988), 29% of untreated animals survived 1 hour of hemorrhage followed by 5 days of resuscitation. All rabbits had apparent injury to their lungs, liver, and intestinal mucosa. Pretreatment with MAb 60.3 significantly attenuated organ injury and allowed 100% survival when the rabbits were subjected to the same protocol. In a more severe model of hemorrhagic shock in rabbits, survival increased from 0% (with 2 hours of shock followed by 5 days of attempted resuscitation) to 71% with MAb 60.3 treatment administered immediately before reperfusion (Vedder et al, 1989). Similar results were obtained in a third study of hemorrhagic shock where two of five monkeys died from 90 minutes of shock and 24 hours of resuscitation, while none died with MAb 60.3 treatment.

Antibodies directed against the α-subunit of the CD11/CD18 adhesion complex are also effective in reducing reperfusion injury. In two studies of canine myocardial I/R, MAb 904 (directed against CD11b) reduced infarct size by 46% (Simpson et al., 1988) and 42% (Simpson et al., 1990). The antibody in these two studies was given at the midpoint of a 90 minute ischemic period. In a third study, hepatic injury was significantly and similarly reduced when a CD11b MAb (Cl-17) was given either prior to ischemia or prior to reperfusion (Jaeschke et al., 1993).

Given that protection is provided by MAbs against CD11/CD18, one might also expect similar benefits from MAbs against a CD11/CD18 counter-receptor on the surface of endothelial cells (i.e., ICAM-1). Two separate studies of myocardial I/R in the monkey (Winquist et al., 1992) and in the cat (Ma et al., 1992) have demonstrated that ICAM-1 MAbs reduce necrosis by 60% (Winquist et al., 1992) and 69% (Ma et al., 1992).

While most of the emphasis on antibody treatment against leukocyte adhesion has focused on prevention of firm adhesion, a few studies have also been performed to investigate the effect of preventing leukocyte rolling with MAbs directed against P-selectin. One such study was performed using the model of arterial occlusion in the rabbit ear described above (Winn et al., 1992). Six hours of occlusion followed by 7 days of reperfusion resulted in 46% necrosis. However, when the P-selectin MAb PB1.3 was added at the onset of reperfusion, tissue necrosis was only 2.7% seven days later, with significantly less edema in the ear. The same MAb was also shown to be effective in a model of myocardial I/R in the cat (Weyrich et al., 1993), where treatment immediately prior to a 270 minute reperfusion period (following 90 minutes of ischemia) significantly attenuated myocardial necrosis by 57%.

These studies with adhesion-blocking antibodies demonstrate that leukocyte-endothelial cell adhesion is a critical step in the overall process of I/R injury. Two lines of evidence suggest that antioxidants should also afford protection in models of I/R. First, interventions directed at scavenging or preventing the production of free radicals are able to attenuate the leukocyte-endothelial cell adhesion elicited by I/R (Del Maestro et al., 1982; Granger et al., 1989; Suzuki et al., 1991b). Second, antioxidants should also afford protection against the oxidants produced by activated leukocytes that are either adherent in postcapillary venules or extravasated and in close proximity to parenchymal cells.

Table 1. Protective Effect of SOD and catalase in I/R injury

Organ or Tissue	*Treatment*	*Reference*
Heart	SOD, catalase	Shlafer et al., 1982
	SOD, catalase	Jolly et al., 1984
	SOD	Burton, 1985
Intestine	SOD	Parks et al., 1982
	SOD	Dalsing et al., 1983
	SOD	Schoenberg et al., 1984
Kidney	SOD	Hansson et al., 1983
	SOD	Paller et al., 1984
Liver	SOD, catalase	Adkison et al., 1986
	SOD, catalase	Simpson et al., 1993
Lung	SOD, catalase	Simpson et al., 1993
Skeletal muscle	SOD	Korthuis et al., 1985
Skin	SOD	Manson et al., 1983
	SOD	Sagi et al., 1986

B. Protection through Antioxidants

Table 1 summarizes some of the literature that demonstrates a protective effect of either SOD or catalase in different models of I/R injury. The results of these studies and others clearly indicate that antioxidant enzymes are effective in reducing the microvascular dysfunction, leukocyte sequestration, and parenchymal cell necrosis elicited by I/R. In most experimental models in which tissues are perfused with whole blood, it is difficult to discern whether SOD and catalase protect by directly interfering with the injury process or do so by limiting the I/R-induced leukosequestration. The former possibility is supported by reports of a protective action of SOD and catalase in isolated hearts perfused with cell-free buffers (Bernier et al., 1986; Omar and McCord, 1990).

VI. SUMMARY

Ischemia-reperfusion tissue injury is associated with the production of reactive oxygen metabolites (ROMs), microvascular dysfunction, leukocyte sequestration, and parenchymal cell necrosis. There is a large body of evidence implicating both ROMs and leukocytes in I/R injury. While the shared protective properties of therapies directed against scavenging ROMs or preventing leukocyte adhesion may be explained by a mechanism in which adherent and extravasated leukocytes produce injury through the production of ROMs (Figure 1), an alternate scheme appears equally feasible (Figure 3). Leukocytes play the leading role in this scenario, with the initial oxidant production following reperfusion primarily serving to upregulate glycoproteins on surfaces of both leukocytes and endothelial cells. This upregulation recruits circulating leukocytes to become adherent and stimulated to release their own oxidants (and proteases) that are ultimately responsible for the majority of the tissue injury. Future work in developing I/R treatments should be directed at assessing the relative input of the mechanisms outlined in Figures 1 and 3.

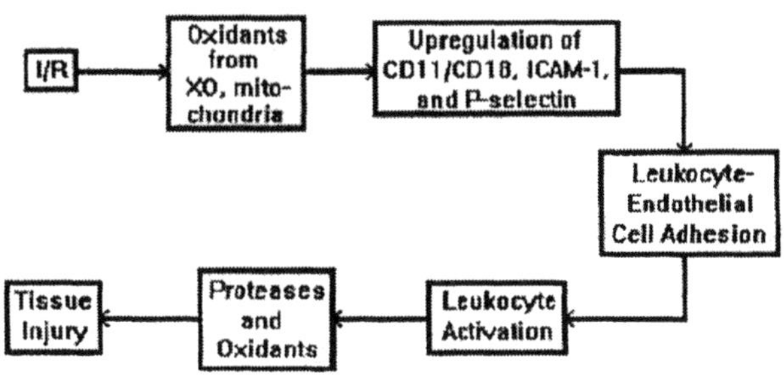

Figure 3. Mechanism where the majority of I/R-induced injury arises from oxidants and proteases produced by adherent leukocytes.

REFERENCES

Adkison, D., Höllwarth, M. E., Benoit, J. N., Parks, D. A., McCord, J. M., & Granger, D. N. (1986). Role of free radicals in ischemia-reperfusion injury to the liver. Acta Physiol. Scand. 548, 101–107.

Barroso-Aranda, J., Schmid-Schönbein, G. W., Zweifach, B. W., & Engler, R. L. (1988). Granulocytes and no-reflow phenomenon in irreversible hemorrhagic shock. Circ. Res. 63, 437–447.

Bernier, M., Hearse, D. J., & Manning, A. S. (1986). Reperfusion-induced arrhythmias and oxygen-derived free radicals: studies with "anti-free radical" interventions and a free radical-generating system in the isolated perfused rat heart. Circ. Res. 58, 331–340.

Bertuglia, S., Colantuoni, A., & Intaglietta, M. (1993). Effect of leukocyte adhesion and microvascular permeability on capillary perfusion during ischemia-reperfusion injury in hamster cheek pouch. Int. J. Microcirc.: Clin. Exp. 13, 13–26.

Burton, K. P. (1985). Superoxide dismutase enhances recovery following myocardial ischemia. Am. J. Physiol. 248, H637–H643.

Carden, D. L., Smith, J. K., & Korthuis, R. J. (1990). Neutrophil-mediated microvascular dysfunction in postischemic canine skeletal muscle: role of granulocyte adherence. Circ. Res. 66, 1436–1444.

Dalsing, M. C., Grosfeld, J. L., Shiffler, M. A., Vane, D. W., Hull, M., Baehner, R. L., & Weber, T. R. (1983). Superoxide dismutase: A cellular protective enzyme in bowel ischemia. J. Surg. Res. 34, 589–596.

Del Maestro, R. F., Planker, M., & Arfors, K.-E. (1982). Evidence for the participation of superoxide anion radical in altering the adhesive interaction between granulocytes and endothelium, in vivo. Int. J. Microcirc.: Clin. Exp. 1, 105–120.

Erlansson, M., Bergqvist, D., Persson, N. H., & Svensjo, E. (1991). Modification of postischemic increase of leukocyte adhesion and vascular permeability in the hamster by Iloprost. Prostaglandins 41(2), 157–168.

Gomoll, A. W., Lekich, R. F., & Grove, R. I. (1991). Efficacy of a monoclonal antibody (MoAb 60.3) in reducing myocardial injury resulting from ischemia/reperfusion in the ferret. J. Cardiovasc. Pharmacol. 17, 873–878.

Granger, D. N. (1988). Role of xanthine oxidase and granulocytes in ischemia-reperfusion injury. Am. J. Physiol. 255, H1269–H1275.

Granger, D. N., Benoit, J. N., Suzuki, M., & Grisham, M. B. (1989). Leukocyte adherence to venular endothelium during ischemia-reperfusion. Am. J. Physiol. 257, G683–G688.

Granger, D. N., Russell, J., Arfors, K. E., Rothlein, R., & Anderson, D. C. (1991). Role of CD11/CD18 and ICAM-1 in ischemia-reperfusion induced leukocyte adherence and emigration in mesenteric venules. FASEB J. 5, A1753.

Granger, D. N., Rutili, G., & McCord, J. M. (1981). Superoxide radicals in feline intestinal ischemia. Gastroenterology 81, 22–29.

Grisham, M. B., Hernandez, L. A., & Granger, D. N. (1986). Xanthine oxidase and neutrophil infiltration in intestinal ischemia. Am. J. Physiol. 251, G567–G574.

Hansson, R., Jonsson, O., Lundstam, S., Pettersson, S., Schersten, T., & Waldenström. (1983). Effects of free radical scavengers on renal circulation after ischaemia in the rabbit. Clin. Sci. 65, 605–610.

Hellberg, P. O. A., Källskog, Ö. T., Öjteg, G., & Wolgast, M. (1990). Peritubular capillary permeability and intravascular RBC aggregation after ischemia: Effects of neutrophils. Am. J. Physiol. 258, F1018–F1025.

Hernandez, L. A., Grisham, M. B., Twohig, B., Arfors, K. E., Harlan, J. M., & Granger, D. N. (1987). Role of neutrophils in ischemia-reperfusion-induced microvascular injury. Am. J. Physiol. 253, H699–H703.

Horgan, M. J., Ge, M., Gu, J., Rothlein, R., & Malik, A. B. (1991). Role of ICAM-1 in neutrophil-mediated lung vascular injury after occlusion and reperfusion. Am. J. Physiol. 261, H1578–H1584.

Jaeschke, H., Farhood, A., Bautista, A. P., Spolarics, Z., Spitzer, J. J., & Smith, C. W. (1993). Functional inactivation of neutrophils with a Mac-1 (CD11b/CD18) monoclonal antibody protects against ischemia-reperfusion injury in rat liver. Hepatology 17, 915–923.

Jaeschke, H., Farhood, A., & Smith, C. W. (1990). Neutrophils contribute to ischemia/reperfusion injury in rat liver in vivo. FASEB J. 4, 3355–3359.

Jolly, S. R., Kane, W. J., Bailie, M. B., Abrams, G. D., and Lucchesi, B. R. (1984). Canine myocardial reperfusion injury: Its reduction by the combined administration of superoxide dismutase and catalase. Circ. Res. 54, 277–285.

Korthuis, R. J., Carden, D. L., Kvietys, P. R., Shepro, D., & Fuseler, J. (1991). Phalloidin attenuates postischemic neutrophil infiltration and increased microvascular permeability. J. Appl. Physiol. 71(4), 1261–1269.

Korthuis, R. J., Granger, D. N., Townsley, M. I., & Taylor, A. E. (1985). The role of oxygen-derived free radicals in ischemia-induced increases in canine skeletal muscle vascular permeability. Circ. Res. 57, 599–609.

Kubes, P., Suzuki, M., & Granger, D. N. (1990). Modulation of PAF-induced leukocyte adherence and increased microvascular permeability. Am. J. Physiol. 259, G859–G864.

Kurose, I., Anderson, D. C., Miyasaki, M., Tamatani, T., Paulson, J. C., Todd, R. F., Rusche, J. R., & Granger, D. N. (1994). Molecular determinants of reperfusion-induced leukocyte adhesion and vascular protein leakage. Circ. Res. 74(2), 336-343.

Kurtel, H., Tso, P., & Granger, D. N. (1992). Granulocyte accumulation in postischemic intestine: role of leukocyte adhesion glycoprotein CD11/CD18. Am. J. Physiol. 262, G878–G882.

Lehr, H. A., Guhlmann, A., Nolte, D., Kepplar, D., & Messmer, K. (1991). Leukotrienes as mediators in ischemia-reperfusion injury in a microcirculation model in the hamster. J. Clin. Invest. 87, 2036–2041.

Ma, X., Lefer, D. J., Lefer, A. M., & Rothlein, R. (1992). Coronary endothelial and cardiac protective effects of a monoclonal antibody to Intercellular Adhesion Molecule-1 in myocardial ischemia and reperfusion. Circulation 86, 937–946.

Manson, P. N., Anthenelli, R. M., Im, M. J., Bulkley, G. B., & Hoopes, J. E. (1983). The role of oxygen-free radicals in ischemic tissue injury in island skin flaps. Ann. Surg. 198, 87–90.

Mullane, K. M., Read, N., Salmon, J. A., & Moncada, S. (1984). Role of leukocytes in acute myocardial infarction in anesthetized dogs: Relationship to myocardial salvage by anti-inflammatory drugs. J. Pharmacol. Exp. Ther. 228(2), 510–522.

Oliver, M. G., Specian, R. D., Perry, M. A., & Granger, D. N. (1991). Morphologic assessment of leukocyte-endothelial cell interactions in mesenteric venules subjected to ischemia and reperfusion. Inflammation 15(5), 331–346.

Omar, B. A., & McCord, J. M. (1990). The cardioprotective effect of Mn-superoxide dismutase is lost at high doses in the postischemic isolated rabbit heart. Free Radic. Biol. Med. 9, 473–478.

Paller, M. S., Hoidal, J. R., & Ferris, T. F. (1984). Oxygen free radicals in ischemic acute renal failure in the rat. J Clin. Invest. 74, 1156–1164.

Parks, D. A., Bulkley, G. B., Granger, D. N., Hamilton, S. R., & McCord, J. M. (1982). Ischemic injury in the cat small intestine: Role of superoxide radicals. Gastroenterology 82, 9–15.

Parks, D. A., & Granger, D. N. (1986). Contributions of ischemia and reperfusion to mucosal lesion formation. Am. J. Physiol. 250, G749–G753.

Perry, M. A., & Granger, D. N. (1992). Leukocyte adhesion in local versus hemorrhage-induced ischemia. Am. J. Physiol. 263, H810–H815.

Romson, J. L., Hook, B. G., Kunkel, S. L., Abrams, G. D., Schork, M. A., & Lucchesi, B. R. (1983). Reduction of the extent of ischemic myocardial injury by neutrophil depletion in the dog. Circulation 67(5), 1016–1023.

Rubin, B. B., Smith, A., Liauw, S., Isenman, D., Romaschin, A. D., & Walker, P. M. (1990). Complement activation and white cell sequestration in postischemic skeletal muscle. Am. J. Physiol. 259, H525–H531.

Sagi, A., Ferder, M., Levens, D., & Strauch, B. (1986). Improved survival of island flaps after prolonged ischemia by perfusion with superoxide dismutase. Plastic Reconstructive Surgery 77, 639–642.

Schmeling, D. J., Caty, M. G., Oldham, K. T., Guice, K. S., & Hinshaw, D. B. (1989). Evidence for neutrophil-related acute lung injury after intestinal ischemia-reperfusion. Surgery 106, 195–202.

Schoenberg, M. H., Muhl, E., Sellin, D., Younes, M., Schildberg, F. W., & Haglund, U. (1984). Posthypotensive generation of superoxide free radicals—Possible role in the pathogenesis of the intestinal mucosal damage. Acta Chir. Scand. 150, 301–309.

Schoenberg, M. H., Poch, B., Younes, M., Schwarz, A., Baczako, K., Lundberg, C., Haglund, U., & Beger, H. G. (1991). Involvement of neutrophils in postischaemic damage to the small intestine. Gut 32, 905–912.

Seekamp, A., Mulligan, M. S., Till, G. O., & Ward, P. A. (1993). Requirements for neutrophil products and L-arginine in ischemia-reperfusion injury. Am. J. Pathol. 142, 1217–1226.

Shiga, Y., Onodera, H., Kogure, K., Yamasaki, Y., Yashima, Y., Syozuhara, H., & Sendo, F. (1991). Neutrophil as a mediator of ischemic edema formation in the brain. Neuroscience Lett. 125, 110–112.

Shlafer, M., Kane, P. F., & Kirsh, M. M. (1982). Superoxide dismutase plus catalase enhances the efficacy of hypothermic cardioplegia to protect the globally ischemic, reperfused heart. J. Thorac. Cardiovasc. Surg. 83, 830–839.

Simpson, P. J., Todd, R. F., Fantone, J. C., Mickelson, J. K., Griffin, J. D., & Lucchesi, B. R. (1988). Reduction of experimental canine myocardial reperfusion injury by a monoclonal antibody (anti-Mo1, anti-CD11b) that inhibits leukocyte adhesion. J. Clin. Invest. 81, 624–629.

Simpson, P. J., Todd, R. F., Mickelson, J. K., Fantone, J. C., Gallagher, K. P., Lee, K. A., Tamura, Y., Cronin, M., & Lucchesi, B. R. (1990). Sustained limitation of myocardial reperfusion injury by a monoclonal antibody that alters leukocyte function. Circulation 81, 226–237.

Simpson, R., Alon, R., Kobzik, L., Valeri, R., Shepro, D., Hechtman, H. B. (1993). Neutrophil and nonneutrophil-mediated injury in intestinal ischemia-reperfusion. Ann. Surg. 218, 444–454.

Smith, S. M., Holm-Rutili, L., Perry, M. A., Grisham, M. B., Arfors, K.-E., Granger, D. N., & Kvietys, P. R. (1987). Role of neutrophils in hemorrhagic shock-induced gastric mucosal injury in the rat. Gastroenterology 93, 466–471.

Suzuki, M., Asako, H., Kubes, P., Jennings, S., Grisham, M. B., & Granger, D. N. (1991a). Neutrophil-derived oxidants promote leukocyte adherence in postcapillary venules. Microvasc. Res. 42, 125–138.

Suzuki, M., Grisham, M. B., & Granger, D. N. (1991b). Leukocyte–endothelial cell adhesive interactions: role of xanthine oxidase-derived oxidants. J. Leuk. Biol. 50, 488–494.

Suzuki. M., Inauen, W., Kvietys, P. R., Grisham, M. B., Meininger, C., Schelling, M. E., Granger, H. J., & Granger, D. N. (1989). Superoxide mediates reperfusion-induced leukocyte-endothelial cell interactions. Am. J. Physiol. 257, H1740–H1745.

Tanaka, M., Brooks, S. E., Richard, V. J., FitzHarris, G. P., Stoler, R. C., Jennings, R. B., Arfors, K.-E., & Reimer, K. A. (1993). Effect of anti-CD18 antibody on myocardial neutrophil accumulation and infarct size after ischemia and reperfusion in dogs. Circulation 87, 526–535.

Vedder, N. B., Fouty, B. W., Winn, R. K., Harlan, J. M., & Rice, C. L. (1989). Role of neutrophils in generalized reperfusion injury associated with resuscitation from shock. Surgery 106, 509–516.

Vedder, N. B., Winn, R. K., Rice, C. L., Chi, E. Y., Arfors, K.-E., & Harlan, J. M. (1988). A monoclonal antibody to the adherence-promoting leukocyte glycoprotein, CD18, reduces organ injury and improves survival from hemorrhagic shock and resuscitation in rabbits. J. Clin. Invest. 81, 939–944.

Vedder, N. B., Winn, R. K., Rice, C. L., Chi, E. Y., Arfors, K.-E., & Harlan, J. M. (1990). Inhibition of leukocyte adherence by anti-CD18 monoclonal antibody attenuates reperfusion injury in the rabbit ear. Proc. Natl. Acad. Sci. USA 87, 2643–2646.

Vejlens, G. (1938). The distribution of leukocytes in the vascular system. Acta Pathol. Microbiol. Scand. 33, 1–239.

Weyrich, A. S., Ma, X., Lefer, D. J., Albertine, K. H., & Lefer, A. M. (1993). In vivo neutralization of P-selectin protects feline heart and endothelium in myocardial ischemia and reperfusion injury. J. Clin. Invest. 91, 2620–2629.

Winn, R. K., Vedder, N. B., Paulson, J. C., & Harlan, J. M. (1992). Monoclonal antibodies to P-selectin are effective in preventing reperfusion injury to rabbit ears. Circulation 86(4), I–80.

Winquist, R., Frei, P., Harrison, P., McFarland, M., Letts, G., Van, G., Andrews, L., Rothlein, R., & Hintze, T. (1990). An anti-CD18 MAb limits infarct size in primates following myocardial ischemia and reperfusion. Circulation 82, III–701 (2785).

Winquist, R. J., Frei, P. P., Letts, L. G., Van, G. Y., Andrews, L. K., Rothlein, R., Dreyer, W. J., Smith, C. W., & Hintze, T. H. (1992). A monoclonal antibody directed against Intercellular Adhesion Molecule-1 limits necrosis and improves function after myocardial ischemia and reperfusion in anesthetized monkeys. J. Vasc. Res. 29, 227.

OXYRADICALS AND ACUTE LUNG INJURY

Jonathan A. Leff, Brooks M. Hybertson, and
John E. Repine

I. INTRODUCTION

The 1969 discovery of the enzymatic function of superoxide dismutase (SOD) by McCord and Fridovich ushered in an explosive era of medical research to determine

Advances in Molecular and Cellular Biology
Volume 25, pages 113-134.

ISBN: 0-7623-0379-4

the role of oxyradicals in health and disease (McCord and Fridovich, 1969). Oxyradicals have been implicated subsequently in many disease processes especially a number of pulmonary diseases (Cross, 1987). This is not totally surprising since lungs receive the entire cardiac output and are, accordingly, abundantly exposed to many blood-borne oxyradicals, such as those derived from phagocytes and circulating xanthine oxidase (XO). Moreover, phagocytes and/or XO may stick preferentially to lung endothelial cells and thereby concentrate their oxyradical generating power in the lung by that mechanism. Furthermore, lungs are also exposed directly to high concentrations of molecular and environmental oxidants in cigarette smoke and polluted air. In this chapter, we review the evidence for the involvement of oxyradicals in acute lung injury (Repine, 1992).

II. MECHANISMS BY WHICH OXYRADICALS CAUSE ACUTE LUNG INJURY

Oxyradicals can injure tissues in numerous ways. Key cellular components are directly susceptible to oxidant attack. Lipids, proteins, and DNA molecules are important targets. Damaging these molecules can have major detrimental effects. For example, when oxyradicals react with lipids, they initiate lipid peroxidation that alters membrane integrity and the cellular microenvironment. Membrane lipids containing unsaturated double bonds are the most susceptible to oxidation. After a lipid peroxide is produced by reacting with a free radical, it propagates by a chain reaction. Membrane proteins involved in ion or other substrate transport may also be perturbed during this process and, as a consequence, damage these important cellular functions. Oxidants damage proteins by several mechanisms: These include (1) direct attack causing conformational change in amino acids, (2) denaturation of proteins, and (3) oxidation which renders proteins more susceptible to hydrolysis. All cells are richly endowed with antioxidant enzymes, but these defenses can be overwhelmed by excessive production of oxyradicals. Furthermore, the localized generation of specific oxidant species may inactivate and thereby undermine the effectiveness of antioxidant enzymes. This may be especially true for hydroxyl radical ($\cdot OH$) for which there are no known enzymatic scavengers. This highly reactive molecule which is formed by the iron catalyzed reaction of superoxide anion ($O_2^{\bullet -}$) and hydrogen peroxide (H_2O_2) is an extremely powerful oxidant. Not surprisingly, oxyradicals readily damage fibroblasts, lung parenchymal cells, red blood cells, endothelial cells, and many other cell types *in vitro* (Simon et al., 1981; Martin et al., 1981; Weiss et al., 1981; Sacks et al., 1978; Weiss, 1980; McCord, 1974).

Oxyradicals can also cause cellular damage indirectly. One example is by potentiating the deleterious effects of elastase released by activated neutrophils. Elastase is a serine protease contained in the azurophilic granules of neutrophils. Originally named for its substrate, elastin, which it readily degrades, elastase also perturbs numerous other substrates including collagen, immunoglobulins, complement components, proteoglycans, clotting factors and fibronectin (Henson et al. 1988; Janoff, 1985). The principal elastase inhibitor in humans is the 52 kDa glycoprotein, α_1-protease inhibitor (α_1Pi, formerly known as α1-antitrypsin) which irreversibly complexes and inactivates elastase (Janoff, 1985). The reaction between elastase and the α1-protease inhibitor is exceptionally fast making elastase-induced tissue injury unlikely when the α1-protease inhibitor is functionally active. Nonetheless, uncomplexed elastase is detectable at sites of tissue injury and therefore it appears that elastase is not always inactivated and may contribute to tissue damage (Janoff, 1985). The plausible explanation is that local antiproteases are inactivated and experimental evidence supports the premise that this mechanism is involved in acute lung injury: First, α_1Pi contains a methionine residue at position 358 in its active site which is extremely sensitive to oxidative attack. Oxidation of methionine by chemical or neutrophil derived oxyradicals inactivates α_1Pi function and increases the half life of elastase 200 fold (Janoff, 1985; Maier et al. 1989; Pryor et al., 1986; Travis and Salveson, 1983). Second, oxidized (inactivated) α_1Pi is increased in lavages from lungs of patients with acute lung injury (Cochrane, et al., 1983). Third, in isolated-perfused rat lungs, perfusion with small doses of elastase does not injure unless the lungs are previously exposed to small, non-injurious concentrations of H_2O_2 (Baird et al., 1986). Finally, oxyradicals may not only inactivate protease inhibitors, but elastase as well, further emphasizing the importance of balance between proteases and antiproteases (Dean et al., 1989). Additional mechanisms by which oxyradicals may damage tissue indirectly include inactivating antioxidant enzymes, mobilizing iron stores from ferritin, activating transcription factors, such as NF-κB, which regulate proinflammatory mediators, generating thromboxane from arachidonic acid and forming peroxynitrite by reacting with nitric oxide (NO) (Repine, 1992; Biemond et al., 1984; Pahl and Banerle, 1994; Tate et al., 1984b).

III. LUNG ANTIOXIDANT DEFENSE SYSTEMS

Given the constant exposure of the lungs to oxidants, via the blood, via the airway, and/or as a consequence of normal or accelerated mitochondrial respiration, it is not surprising that an intricate antioxidant defense system has evolved to ensure an appropriate balance between oxidants (which may also have beneficial roles) and antioxidants (Heffner and Repine, 1989). The extent and delicate nature of this balance is just now being appreciated. The beneficial effect of oxyradicals now encompasses their role as mediators as well as their bactericidal contributions. This system includes both intracellular and extracellular components which reduce formation of oxyradicals, increase the metabolism of oxyradicals to less toxic forms, localize the reactions of oxyradicals to less vital cellular locations and/or repair oxidized targets (Heffner and Repine, 1989; Sies, 1987). For example, mitochondria compartmentalize their reduction of molecular oxygen which localizes the very small concomitant leak of oxyradicals which occurs under normal circumstances of cellular metabolism. When this "leak" is exaggerated under conditions of hyperoxia, mitochondrial forms of SOD and catalase scavenge the leaked oxyradicals (Guidot et al., 1995). Special molecules in lung cells and extracellular fluids regulate tightly the amount of free iron and copper and prevent their participation in oxidant potentiating reactions (Heffner and Repine, 1989). These metals catalyze Haber-Weiss mediated generation of hydroxyl radical, facilitating lipid peroxidation, and accelerate nonenzymatic oxidation of epinephrine and other molecules.

The primary antioxidant enzymes in lung cells include catalase, superoxide dismutase (ECSOD, Cu,Zn-SOD and MnSOD), glucose-6-phosphate dehydrogenase, and the enzymes of the glutathione redox cycle, glutathione peroxidase and glutathione reductase (Heffner and Repine, 1989; Sies, 1987). Lung cells also contain two lipid soluble antioxidants which are vitamin E (α-tocopherol) and beta-carotene. Vitamin E, by virtue of its lipid solubility, intercalates into lung cell membranes where it interrupts the cascade of lipid peroxidation. Vitamin E is also present in extracellular fluids and plasma. Beta-carotene also accumulates in the membranes where it presumably quenches singlet oxygen (Heffner and Repine, 1989; Sies, 1987).

Lung cells are also endowed with several water soluble antioxidants, including vitamin C, glutathione, uric acid, and taurine, which may decrease the toxicity of oxidants. Additionally, certain high molecular weight antioxidants, such as tracheobronchial mucus and albumin, may act as suicide substrates preventing damage to other more vital cellular components. Since the alveolar space is often exposed to inhaled and macrophage-derived oxidants, the antioxidant properties of the epithelial lining fluid (ELF) may have evolved as a compensatory mechanism. For example, for unknown reasons, ELF contains approximately 100 times the reduced glutathione (GSH) concentrations of plasma (Cantin et al., 1987). In addition, ELF has a constitutive catalase activity which ostensibly limits H_2O_2-induced damage (Yoo et al., 1994).

The ability to survive an oxidant insult may depend on repair of damaged tissue. Cellular systems repair DNA strand breaks as well as peroxidized membrane phospholipids (Schraufstatter et al., 1986; Berger, 1985; Van Kuijk et al., 1987). If oxidant-induced damage is too severe to repair, then tissue is replaced by increased proliferation of cells (Thet et al., 1984). These repair processes may inadvertently complicate the recovery of patients who survive acute lung injury.

IV. EVIDENCE FOR OXYRADICAL-INDUCED ACUTE LUNG INJURY IN PATIENTS

The acute respiratory distress syndrome (ARDS) is a defined, inflammatory process that alters lung function severely and ultimately leads to a highly-fatal respiratory failure. For unknown reasons, ARDS is a common response to a diverse variety of insults (Table 1) (Fowler et al., 1983). Pneumonia or toxic inhalation are

Table 1. Risk Factors Associated With Development of the Acute Respiratory Distress Syndrome.

▪ Sepsis
▪ Aspiration (gastric, near drowning)
▪ Trauma
▪ Shock (any cause)
▪ Fat embolism
▪ Drugs (ASA, narcotics [heroine, methadone, propoxyphene], chloroform, colchicine, barbiturates, phosgene, ozone, tocolytics [terbutaline, IL-2, protamine])
▪ Hypertransfusion (>10 units blood products)
▪ Pneumonia
▪ Hyperoxia
▪ Lung contusion
▪ Toxic inhalation
▪ CNS disease (neurogenic pulmonary edema)
▪ Pancreatitis
▪ Uremia
▪ Cardiopulmonary bypass
▪ Burn

the primary airway insults, while sepsis, pancreatitis and trauma are the principal blood-borne insults that appear causative. ARDS is not a reportable disease, but over 150,000 cases may occur yearly using strict diagnostic criteria and many more borderline cases certainly occur. The mortality rate for ARDS patients has remained unchanged (30–60%) largely because of our ignorance of the precise initiating, perpetuating, and terminating mechanisms of lung injury and the lack of an effective treatment (Bernard et al., 1987). The physiological hallmarks of ARDS are pulmonary edema involving capillary leak of protein-rich fluid, stiff lungs with reduced compliance, and severe hypoxemia usually requiring mechanical ventilation. The pulmonary edema is noncardiogenic and caused by leaky pulmonary blood vessels rather than cardiac dysfunction. Lungs from ARDS patients are congested and histologically reveal massive infiltration by neutrophils, interstitial edema and alveolar cellular debris. Several days later lung pathology is characterized by hyaline membrane formation, and subsequently by fibrosis.

Although the precise mechanisms responsible for acute lung injury are not known, many lines of evidence suggest a crucial role for neutrophil-derived oxyradicals. The "neutrophil" hypothesis of ARDS arose in a rather interesting fashion. Initially, an important observation was made that patients undergoing hemodialysis developed complement activation, leukopenia, pulmonary vascular leukocyte sequestration, and pulmonary dysfunction (Craddock et al., 1977; Craddock et al., 1977). Although the hemodialysis-induced pulmonary dysfunction was mild and transient, the notion arose that various inciting stimuli, perhaps in combination ("two-hit" theory), lead to accumulation of neutrophils in the lung, neutrophil activation and release of damaging neutrophil-derived toxins (Leff and Repine, 1993). A large body of clinical evidence now supports this general premise (Tate and Repine, 1983): first, bronchoalveolar lavage and lung biopsies of ARDS patients are massively infiltrated with neutrophils. While it might be argued that the neutrophil influx in ARDS patients is simply a response to injury, neutrophils are often present at very early time points—a finding consistent with the premise that they participate in injury (Gadek, 1992). Second, neutrophil products (e.g., elastase, myeloperoxidase [MPO], elastase-$\alpha 1$Pi complexes, H_2O_2) accumulate in lung lavages, plasma and the exhaled breath of ARDS patients suggesting that the neutrophils are active (Gadek, 1992; Zheutlin et al., 1986; Rocker et al., 1989; Baldwin et al., 1986). Third, even when ARDS occurs, albeit rarely, in neutropenic patients, it worsens when blood neutrophils return (Laufe et al., 1986; Ognibene et al., 1986; Rinaldo and Broovetz, 1985). Fourth, enhanced neutrophil stimulating molecules, such as C5a, tumor necrosis factor (TNF), interleukin-1 (IL-1) and IL-8, exist in the plasma and lung lavages of ARDS patients (Hammerschmidt et al., 1980; Parsons et al., 1989; Hosea et al., 1980). Fifth, abnormalities in the activity of recoverable blood neutrophils in ARDS patients indicate that neutrophils may either be activated or exhausted following activation (Zimmerman et al., 1983; Parsons et al., 1990; Martin et al., 1991). The latter interpretation is, however, complicated by the selection process involved in sampling and testing neutrophils from peripheral blood when the most relevant neutrophils may already reside in the lung.

The participation of neutrophils in the pathogenesis of ARDS has also focused attention on oxyradicals and neutrophil elastase. Although substantial evidence indicates that neutrophil elastases are important, the soundness of this conclusion has been hindered by the lack of effective elastase inhibitors and markers of proteolytic damage (Fowler et al., 1982; McGuire et al., 1982). Evidence implicating oxyradicals, presumably neutrophil-derived, in the pathogenesis of ARDS include studies that show that α_1Pi has been oxidatively inactivated in ARDS patients (Cochrane et al., 1983; Fowler et al., 1982; McGuire et al., 1982). Additionally, elevated levels of H_2O_2 exist in the condensed breath of patients with ARDS (Baldwin et al., 1986; Sznajder et al., 1989; Wilson et al., 1992) and rats given IL-1 intratracheally (Leff et al., 1993b; Leff et al., 1994d). ARDS patients also have decreased GSH levels in their erythrocytes (RBC) and their epithelial lining fluid, indicative of consumption of oxidative defenses with conversion of GSH to GSSG (Bernard et al., 1989). Augmenting GSH by N-acetylcysteine (NAC) or procysteine has corrected GSH abnormalities and reduced organ dysfunction in some patients with ARDS (Bernard et al., 1989; Pacht et al., 1991; Bunnell and Pacht, 1993). NAC treatment has also supplemented antioxidant defenses in a variety of *in vivo* and *in vitro* model systems including rats exposed to paraquat (Hybertson et al., 1995b) or hyperoxia (Patterson et al., 1985) and sheep treated with endotoxin (Bernard et al., 1984). Interestingly, but for unknown reasons, MnSOD and catalase are increased in the serum of septic patients and septic patients who develop ARDS (Leff et al., 1993a). These elevations could simply reflect release of antioxidants from damaged cells or be related to the adaptive tolerance process such as that which occurs in a variety of systems following cytokine treatment where induction of MnSOD is prominent (Repine, 1994).

Oxygen toxicity complicates the course of many intensive care unit patients with acute lung injury who require mechanical delivery of high concentrations of oxygen (Barber et al., 1970; Deneke and Fanburg, 1980). Injury most likely occurs after continuous exposure to hyperoxia for 48 to 72 hours, but it may be hastened by antecedent or evolving lung damage. Exposure to hyperoxia undoubtedly increases production of toxic oxyradicals and this toxicity is most likely a consequence of the effects of oxyradicals on resident lung cells. For example, when exposed to high oxygen tensions, alveolar macrophages release chemotactic and adherence factors for neutrophils (Harada et al., 1984; Bowman et al., 1983). A subsequent influx of neutrophils then occurs and most likely contributes additional injury (Shasby et al., 1982b; Parrish et al., 1984). The earliest injury may depend on the direct toxicity of oxygen to cells. Hyperoxia is toxic to pulmonary endothelial and epithelial cells as well both *in vivo* and *in vitro*. Several animal studies suggest that mitochondria are a predominant source of oxyradicals following continuous exposure to hyperoxia (McCord and Omar, 1993; Turrens et al., 1982; Yusa et al., 1984). Mitochondrial respiration is also a significant source of $O_2^{\bullet -}$ generation in yeast exposed to hyperoxia *in vivo* (Guidot et al., 1993). Other investigations confirm that oxyradicals mediate lung injury during hyperoxia. Hyperoxia potentiates lung damage from agents whose toxicity is oxyradical mediated (Southorn and Powis, 1988). In addition, rats deficient in endogenous antioxidants (SOD or GSH) develop worse lung injury and have diminished survival during hyperoxia (Frank et al., 1981; Deneke et al., 1985). Treatments with endotoxin, IL-1 or TNF or exposure to hypoxia, ozone, or 85% oxygen increase lung antioxidant enzyme activities and protect rats during hyperoxia exposure (Iqbal et al., 1989; Frank et al., 1980; White et al., 1988; Jackson and Frank, 1984). Exogenous antioxidant treatment is protective during hyperoxia exposure. Insufflating RBC or liposome-encapsulated SOD or catalase intratracheally decreases oxygen toxicity in rats (Crapo and Tierney, 1974; van Asbeck et al., 1985; Padmanabhan et al., 1985). Similarly, augmenting endothelial cell SOD (by liposomes) decreases hyperoxia-induced injury *in vitro* (Freeman et al., 1983).

V. EVIDENCE FOR OXYRADICAL-INDUCED ACUTE LUNG INJURY IN EXPERIMENTAL SYSTEMS

Innumerable studies using intact animals, isolated-perfused lung preparations, and cultured lung cells suggest convincingly that oxyradicals contribute to acute lung injury. The comprehensive story using the potent neutrophil stimulant, phorbol myristate acetate (PMA), is a representative and instructive example. In 1974, it was discovered that a soluble chemical, PMA, the essence of croton oil, could stimulate the oxidative metabolism of neutrophils *in vitro* (Repine et al., 1974). This finding showed that neutrophils could respond metabolically to soluble stimuli as well as to particle (such as bacteria) phagocytosis. It was also recognized that PMA powerfully stimulated oxyradical production by normal neutrophils, but not neutrophils from patients with chronic granulomatous disease (CGD), and that PMA also increased neutrophil adherence *in vitro* (Repine et al., 1974). Therefore, PMA appeared to be an ideal agent for increasing the adherence of neutrophils to vascular endothelial cells and stimulating production of oxyradicals by neutrophils *in vivo*. Subsequently, injecting rabbits with PMA intravenously caused an acute, edematous lung injury manifested by decreases in peripheral blood neutrophils, and increases in lung to body weight ratios, lung lavage albumin concentrations, and histological abnormalities (Figure 1) (Shasby et al., 1982a). Furthermore, PMA-induced lung leak was decreased by depleting neutrophils or treating with the highly-permeant, oxyradical scavenger dimethylthiourea (DMTU) (Shasby et al., 1982b). Additional studies revealed that acute edematous injury also occurred in isolated rat lungs perfused with human neutrophils and PMA, but did not occur when CGD neutrophils which lack the ability to generate oxygen radicals were substituted for normal neutrophils (Shasby et al., 1982a). Taken *en toto*, these experi-

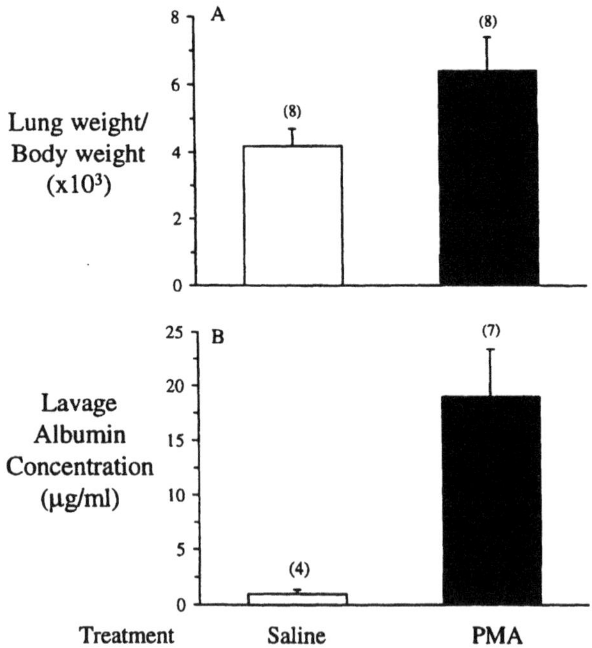

Figure 1. Intravenous injection of PMA in rabbits rapidly (4h) increased lung weight/body weight ratios (**A**) and lavage albumin concentrations (**B**) compared to saline-treated rabbits. Adapted from Shasby et al. (1982a).

ments provided compelling evidence that neutrophils can directly and independently cause lung injury and that neutrophil-derived oxyradicals can be injurious. Further studies showed that PMA-stimulated neutrophils lysed cultured endothelial cells and increased permeability across endothelial cell monolayers, and again, that injury did not occur with CGD neutrophils or after adding oxyradical scavengers (Martin, 1984; Shasby et al., 1983).

Administering other neutrophil activators also causes acute lung injury. For instance, injecting cobra venom factor (CVF) intravenously into rats activates the complement cascade and initiates lung neutrophil sequestration and acute lung injury (Till et al., 1982). Again, lung injury is decreased by neutrophil depletion or antioxidant treatment indicating that neutrophil-derived oxygen metabolites mediate injury. Instilling formylated tripeptide formyl-norleucyl-leucyl-phenylalanine (FNLP) intratracheally also causes neutrophil influx and leak of intravascular albumin into lung tissue in hamsters (Figure 2) (Leff et al., 1992). Treatment with DMSO—an ·OH scavenger—decreased both lavage neutrophil influx and lung

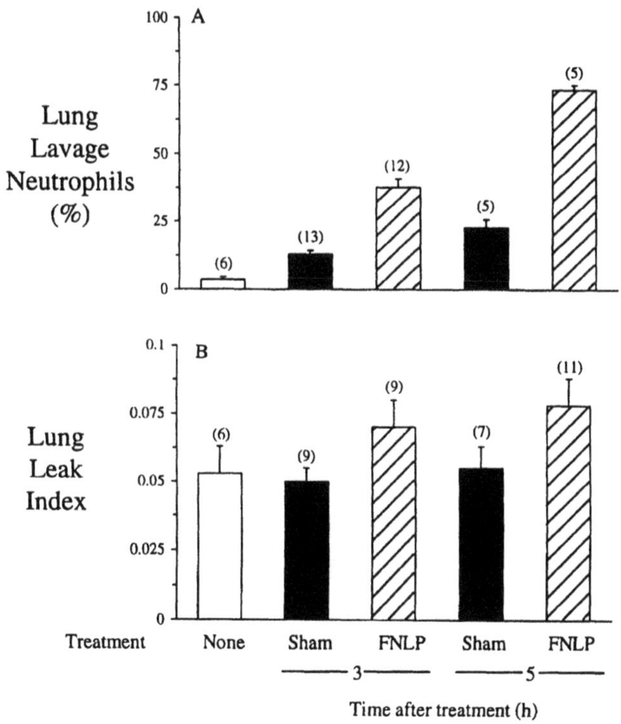

Figure 2. Intratracheal instillation of FNLP (80 μg) increased lung lavage neutrophil percentage (**A**) and lung leak index (**B**) in hamsters compared to sham-treated hamsters. Adapted from Leff et al. (1992).

leak (Figure 3). However, because very low concentrations of DMSO (0.01%) also inhibited neutrophil chemotaxis *in vitro* it is unclear if DMSO protected by blocking neutrophil influx, by scavenging OH and/or some other unidentified mechanism. Another mediator of oxyradical-induced acute lung injury is endotoxin that is elevated in the plasma of patients with acute lung injury (Parsons et al., 1989). Endotoxin causes pulmonary neutrophil influx and lung injury that appears to be oxidant-mediated because it is prevented by catalase treatment (Milligan et al., 1988). Endotoxin also damages endothelial cells (EC) by increasing cellular generation of oxyradicals (Brigham et al., 1987).

Circulating XO may be another source of oxyradicals that cause acute lung injury (Repine et al., 1987). For example, acute paraquat-mediated lung injury in rats is prevented by inhibiting XO - an observation suggesting that XO-derived oxyradicals mediate paraquat-induced lung injury (Waintrub et al., 1990). XO-derived oxyradicals could cause injury indirectly by generating chemotaxins for neutrophils (Repine et al., 1987; Petrone et al., 1980). Blood levels of XO, xanthine and hypoxanthine are elevated in patients with acute lung injury compared to patients without lung injury (Grum et al., 1987). Perfusing

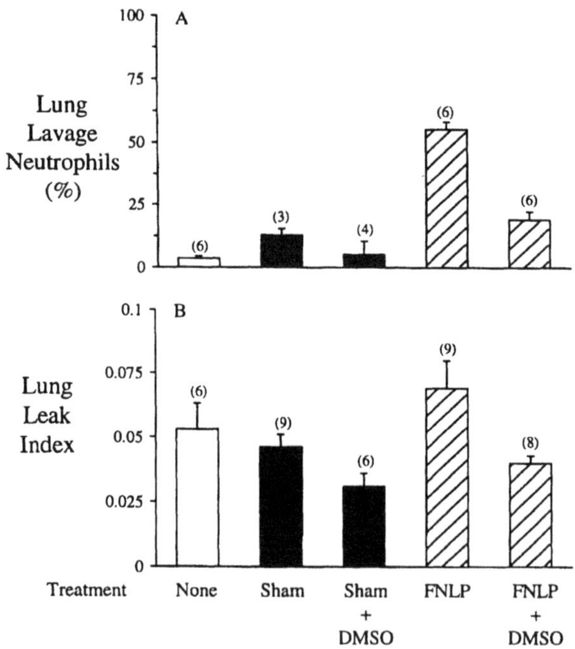

Figure 3. Pretreatment with dimethylsulfoxide (DMSO) attenuated the increases at 3h in both lavage neutrophil. Percent (**A**) and lung leak index (**B**) that occur in hamsters given FNLP intratracheally compared to sham-treated hamsters. Adapted from Leff et al. (1992).

isolated-perfused rabbit lungs with XO increases perfusion pressures, thromboxane levels, and acute edematous injury (Tate et al., 1984a). Rats exposed to a 30% skin burn develop lung leak that is decreased by XO inactivation (Burton et al., 1989). Blood XO levels dramatically rise, and lung neutrophil influx and lung injury occur after reperfusion of ischemic rat intestines (Terada et al., 1992). These two phenomenon appear related since inhibiting XO reduces the increase in blood XO activity as well as lung neutrophil influx and injury after ischemia and reperfusion (I/R). Furthermore, injecting XO exogenously increases plasma XO activity and lung neutrophil sequestration, but not lung injury. Conversely, injecting XO antisera decreased plasma XO activity and lung neutrophil sequestration in rats subjected to intestinal I/R and hyperoxia (Terada et al., 1992). These studies suggest that circulating XO may, in some cases, mediate lung neutrophil recruitment without causing lung injury. Sequestered neutrophils may then be activated by other factors that lead to lung injury. There is also evolving support for the possibility that circulating XO may adhere to vascular endothelium localizing its oxidant insult and/or neutrophil recruitment to lung EC (Adachi et al., 1993).

Cellular XO is also a potential source of oxyradicals. Endotoxin damage to EC is decreased by inhibiting XO (Brigham et al., 1987). More direct measurements in EC subjected to anoxia/reoxygenation, utilizing electron paramagnetic resonance, indicate that intracellular generation of ·OH is decreased by treatment with the XO inhibitor, oxypurinol (Zweier et al., 1988). In addition to neutrophils and XO, mononuclear cells (alveolar macrophages and monocytes), mitochondria, arachidonate metabolism, and autoxidation of catecholamines may also produce oxyradicals (McCord and Omar, 1993).

Cytokines are likely contributors to lung and other tissue injury (Dinarello and Wolff, 1993). IL-1 and TNF levels are increased in the lung lavages of patients with ARDS. In addition, alveolar macrophages obtained from patients with ARDS make more IL-1 than macrophages collected from control patients without ARDS (Siler et al., 1989; Suter et al., 1992; Jacobs et al., 1989). Instilling IL-1 intratracheally causes neutrophil influx into the alveoli of steers, rats and hamsters (Repine, 1994; Heidel et al., 1990; Ulich et al., 1991; Patton et al., 1995).

We further defined the role of IL-1 in acute lung injury by insufflating IL-1 into the airway of rats. Administering IL-1 intratracheally caused a dose-dependent leak of ^{125}I- labelled albumin (increased permeability) into the lung as well as a neutrophil influx into the alveolar space (as measured by lung lavage analysis) (Figure 4) (Leff, 1994d; Repine, 1994). IL-1-dependent lung lavage neutrophil accumulation and lung leak were prevented by pretreating the rats with the IL-1 receptor antagonist (IL-1ra), by heating the IL-1 before instillation, and by administering the IL-1 intravenously instead of intratracheally (Figure 5) (Leff, 1994d; Repine, 1994; Leff et al., 1994e). These findings suggested that IL-1, via a receptor-mediated mechanism in the airway, was responsible for lung neutrophil influx and lung leak. Since IL-1 is not intrinsically a potent neutrophil

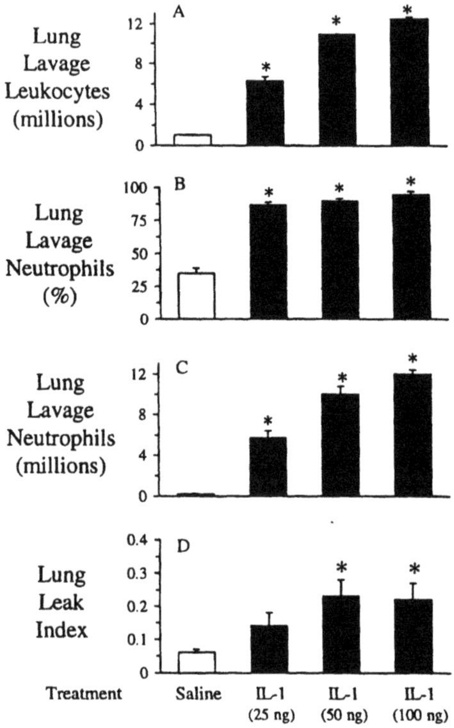

Figure 4. Intratracheal insufflation of increasing concentrations of IL-1 progressively increased lavages leukocytes (**A**), lavage neurophil percentages (**B**), lavage neurophils (**C**) and lung leak index (**D**) compared to saline-treated rats; * $p<0.05$ compared to saline-treated rats. Adapted from Leff et al. (1994d).

chemotaxin or stimulant of oxyradical production *in vitro*, interaction with airway cells and elaboration of additional mediators appeared essential. We found that intratracheal insufflation of IL-1 in rats increased lung lavage cytokine-induced neutrophil chemoattractant (CINC) levels and that co-treatment with anti-CINC antibody decreased IL-1-induced lavage neutrophil influx and lung leak (Figure 6) (Koh et al., 1995a; Koh et al., 1995b). CINC is a potent rat neutrophil chemotaxin which is functionally and structurally related to human IL-8 and MGSA/gro cytokines (Miller and Krangel, 1992). Similarly, levels of mRNA for macrophage inflammatory protein-2 (MIP-2) and KC protein (CINC) were elevated in both lavage cells and tracheal homogenates from rats treated with endotoxin (Huang et al., 1992).

We also studied the phenomenon of IL-1-induced acute lung injury in an isolated, buffer-perfused rat lung which is devoid of circulating blood cells, nervous innervation, and other complicating factors. In these studies, human neutrophil

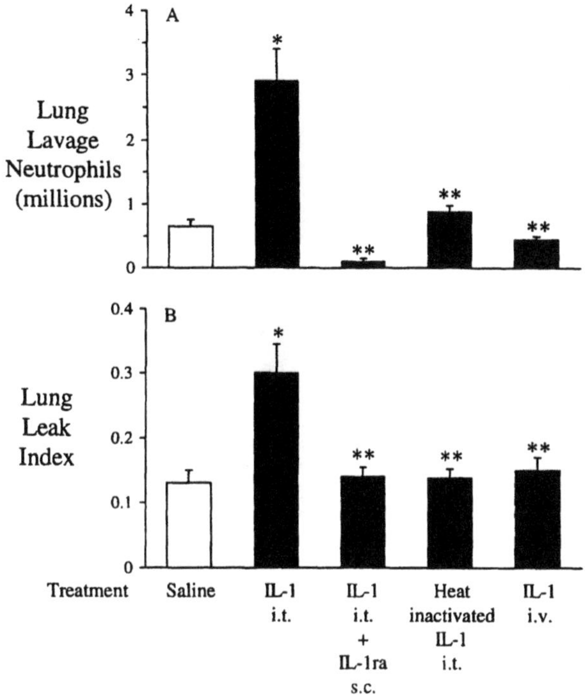

Figure 5. Intratracheal insufflation of IL-1 (50 ng) increased lung lavage neutrophils (**A**) and lung leak index (**B**) compared to saline treated rats. Both IL-1-induced increases were attenuated by pretreatment (t=-30 min) with IL-1 receptor antagonist (IL-1ra), by using heat-inactivated IL-1 for insufflation, or by giving IL-1 intravenously instead of intratracheally; * $p<0.05$ compared to saline-treated rats; ** $p<0.05$ compared to IL-1 treated rats. Adapted from Leff et al. (1994d).

perfused rat lungs were injured after administering IL-1 intratracheally but not injured when neutrophils were not added, or when the neutrophils were heated to inactivate the NADPH-oxidase complex ($O_2^{\bullet -}$ production) but not chemotaxis or adherence *in vitro*. These studies suggested that neutrophil-derived oxyradicals were responsible for injury in isolated rat lungs given IL-1 intratracheally (Guidot et al., 1994a).

Leukotrienes (e.g., LTB_4) may mediate IL-8-induced lung injury. Leukotrienes are increased in lung lavages from patients with acute lung injury (Stephenson et al., 1988; Seeger et al., 1991). Moreover, in isolated, buffer-perfused rat lungs, inhibiting 5-lipoxygenase activity (with Zileuton) decreased leukotriene synthesis and protected rat lungs against injury when IL-8 is administered intratracheally and the lungs are perfused with human neutrophils (Guidot et al., 1994b).

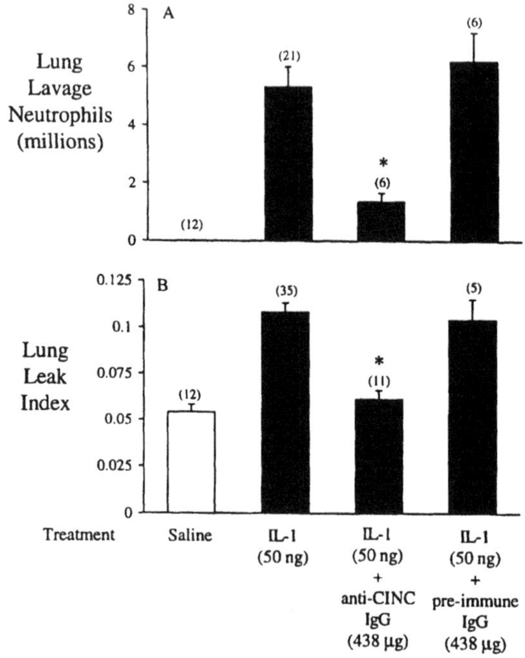

Figure 6. Intratracheal insufflation of anti-CINC antibody along with IL-1 attenuated IL-1-induced increases in lavage neutrophils (**A**) and lung leak (**B**) in rats. This protection was not afforded by co-treatment with pre-immune IgG; *p.05 compared to IL-1-treated rats. Adapted from Koh et al. 1995b.

We examined the participation of neutrophils in IL-1-induced lung injury in intact rats by depleting circulating neutrophils using vinblastine treatment (Leff et al., 1994f; Lemanske et al., 1982). We then challenged these rats by giving IL-1 intratracheally. We found that vinblastine-treated, neutrophil-depleted rats resisted IL-1-induced lung injury whereas vinblastine-treated, neutrophil-replete rats developed lung injury.

To further define the participation of neutrophils in IL-1-induced lung injury, we next assessed the effect of liposome-entrapped prostaglandin E_1 (PGE_1) treatment on IL-1-induced lung injury. PGE_1 increases neutrophil cAMP levels and, as a result, decreases neutrophil adherence, chemotaxis, $O_2^{\bullet -}$ production, and protease release *in vitro*. Liposomes target neutrophils effectively and, accordingly, may enhance the local concentrations of PGE_1. We found that liposomal PGE_1 treatment decreased lung leak and lung neutrophil influx in rats given IL-1 intratracheally (Leff et al., 1994c).

To assess the contribution of oxyradicals to IL-1-induced lung injury, we measured lung levels of oxidized glutathione (GSSG) and breath H_2O_2 levels.

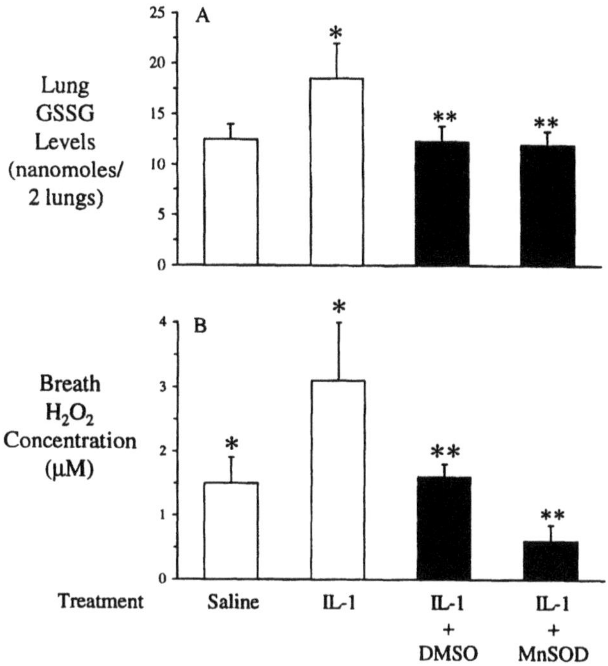

Figure 7. Rats treated with IL-1 by intratracheal insufflation had increased lung GSSG levels (**A**) and breath H_2O_2 levels (**B**) compared to rats treated with saline. Pretreatment with DMSO or MnSOD attenuated the IL-1-induced increases in both lung GSSG levels (**A**) and breath H_2O_2 levels (**B**); * p.05 compared to saline-treated rats, ** $p<0.05$ compared to IL-1-treated rats. Adapted from Leff et al. (1994d).

We found that lungs from rats given IL-1 intratracheally had elevated lung GSSG levels compared to sham treated rats. In addition, condensed breath H_2O_2 concentrations were elevated in rats given IL-1 intratracheally compared to control rats given saline intratracheally (Figure 7). Elevations in these markers of oxidant activity suggested that IL-1-induced lung injury involves oxyradicals (Repine, 1994) and this possibility was confirmed when we pretreated rats with several antioxidants before challenge with IL-1 intratracheally. Pretreatment with DMSO or manganese superoxide dismutase (MnSOD) decreased lung GSSG and expired H_2O_2 levels in rats given IL-1 intratracheally (Repine, 1994). Furthermore, DMSO and MnSOD treatment decreased lung leak and lung lavage neutrophil influx in rats given IL-1 (Leff et al., 1994d; Repine, 1994; Figure 8). Similarly, inhalation and pulmonary deposition of supercritical fluid aerosolized vitamin E prevented lung injury, but vitamin E deficiency exacerbated lung injury in rats given IL-1 intratracheally (Hybertscn et al., 1995a).

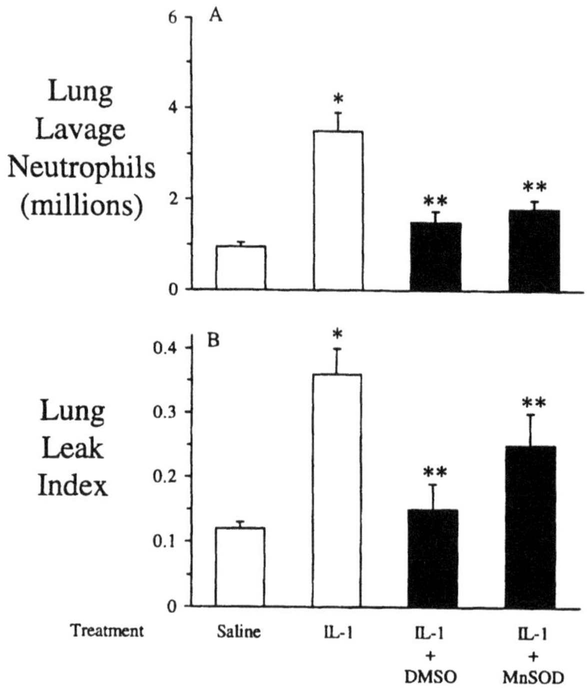

Figure 8. Rats treated with IL-1 by intratracheal insufflation had increased lung lavage neutrophils (**A**) and lung leak index (**B**) compared to rats treated with saline. Pretreatment with DMSO or MnSOD attenuated the IL-1-induced increases in both lavage neutrophils (**A**) and lung leak index (**B**); * p.05 compared to saline treated rats, ** $p<0.05$ compared to IL-1-treated rats. Adapted from Leff et al. (1994d).

VI. FUTURE CHALLENGES

Establishing and understanding the contribution of oxyradicals to lung injury would be enhanced remarkably by direct measurement of oxyradicals. These measurements would indicate if and when oxyradicals are being generated and their relationship to evolving abnormalities in these sick patients. These measurements would also indicate if interventions reduce oxyradicals and associated abnormalities. A second challenge is to treat these individuals with effective antioxidants in appropriate doses at proper times. These are not insignificant aspects since administering excessive antioxidants may actually worsen oxidative damage (Nelson et al., 1994). A third challenge will be to define and preserve the beneficial effects of oxyradicals which include bactericidal functions, termination of lipid peroxidation, modulation of vascular tone and undoubtedly a vast of array of unrecognized effects.

VII. CONCLUSION

The world awaits the successful treatment of any disease with an antioxidant strategy. We are hopeful that effective treatment of pulmonary disorders will occur soon and confirm the enormous amount of evidence predicting this possibility.

REFERENCES

Adachi, T., Fukushima, T., Yoshiko, U., & Kazuyuki, H. (1993). Binding of human xanthine oxidase to sulphated glycosaminoglycans on the endothelial-cell surface. Biochem. J. 289, 523–527.

Baird, B.R., Cheronis, J.C., Sandhaus, R.A., Berger, E.A., White, C.W., & Repine, J.E. (1986). Oxygen metabolites and neutrophil elastase synergistically cause edematous injury in isolated rat lungs. J. Appl. Physiol. 61, 2224–2229.

Baldwin, S.R., Simon, R.H., Grum, C.H., Ketai, L.H., Boxer, L.A., & Devall, L.J. (1986). Oxidant activity in expired breath of patients with adult respiratory distress syndrome. Lancet 1, 11–14.

Barber, R.E., Lee, J., & Hamilton, W.K. (1970). Oxygen toxicity in man: A prospective study in patients with irreversible brain damage. N. Engl. J. Med. 283, 1478–1484.

Berger, N.A. (1985). Poly (ADP-ribose) in the cellular responses of DNA damage. Radiat. Res. 101, 4–15.

Bernard, G.R., Lucht, W.D., Niedermeyer, M.E., Snapper, J.R., Ogletree, M.L., & Brigham, K.L. (1984). Effect on N-acetylcysteine on the pulmonary response to endotoxin in the awake sheep and upon *in vitro* granulocyte function. J. Clin. Invest. 73, 1772–1784.

Bernard, G.R., Luce, J.M., Sprung, C.L., Rinaldo, J.E., Tate, R.M., Sibbald, W.J., Kariman, K., Higgins, S., Bradley, R., Metz, C.A., Harris, T.R., & Brigham, K.L. (1987). High-dose corticosteroids in patients with the adult respiratory distress syndrome. N. Engl. J. Med. 317, 1565–1570.

Bernard, G.R., Swindell, B.B., Meredith, M.J., Carroll, F.E., & Higgins, S.B. (1989). Glutathione (GSH) repletion by N-acetylcysteine in patients with the Adult Respiratory Distress Syndrome. Am. Rev. Resp. Dis. 139, A221. (Abstract)

Biemond, P., van Eijk, H.G., Swaak, A.J.G., & Koster, J.F. (1984). Iron mobilization from ferritin bysuperoxide derived from stimulated polymorphonuclear leukocytes. Possible mechanism in inflammation diseases. J. Clin. Invest. 73, 1576–1579.

Bowman, C.M., Harada, R.N., & Repine, J.E. (1983). Hyperoxia stimulates alveolar macrophages to produce and release a factor which increases neutrophil adherence. Inflammation 7, 331–338.

Brigham, K.L., Meyrick, B., Berry, L.C., & Repine, J.E. (1987). Antioxidants protect cultured bovine lung endothelial cells from injury by endotoxin. J. Appl. Physiol. 63, 840–850.

Bunnell, E. & Pacht, E.R. (1993). Oxidized glutathione is increased in the alveolar fluid of patients with the adult respiratory distress syndrome. Am. Rev. Resp. Dis. 148, 1174–1178.

Burton, L.K., Patt, A., Velasco, S.E., Terada, L.S., Grosso, M.A., Mulvin, D.W., Harken, A.H., & Repine, J.E.(1989). Xanthine oxidase increases in blood and contributes to neutrophil recruitment and oxidative edematous injury (ARDS) in lungs of rats subjected to skin burn. Clin. Res. 37, 145A. (Abstract)

Cantin, A.M., North, S.L., Hubbard, R.C., & Crystal, R.G. (1987). Normal epithelial lining fluid contains high levels of glutathione. J. Appl. Physiol. 63, 152-157.

Cochrane, C.G., Spragg, R.G., & Revak, S.D. (1983). Pathogenesis ofthe adult respiratory distress syndrome: evidence of oxidant activity in bronchoalveolar lavage fluid. J. Clin. Invest. 71, 754–758.

Craddock, P.R., Fehr, J., Brigham, K.L., Kronenberg, R.S., & Jacob, H.S. (1977). Complement and leukocyte-mediated pulmonary dysfunction in hemodialysis. N. Engl. J. Med. 296, 769–774.

Craddock, P.R., Fehr, J., Dalmasso, A.P., Brigham, K.L., & Jacob, H.S. (1977). Hemodialysis leukopenia. Pulmonary vascular leukostasis resulting from complement activation by dialyzer cellophane membranes. J. Clin. Invest. 59, 879–888.

Cross, C.E. (1987). Oxygen radicals and human disease. Ann. Intern. Med. 107, 526–545.

Crapo, J.D., & Tierney, D.F. (1974). Superoxide dismutase and pulmonary oxygen toxicity. Am. J. Physiol. 226, 1401–1407.

Dean, R.T., Nick, H.P., & Schnebli, H.P. (1989). Free radicals inactivate human neutrophil elastase and its inhibitors with comparable efficiency. Biochem. Biophys. Res. Commun. 159, 821–827.

Deneke, S.M., & Fanburg, B.L. (1980). Normobaric oxygen toxicity of the lung. N. Engl. J. Med. 303, 76–86.

Deneke, S.M., Lynch, B.A., & Sanberg, B.L. (1985). Transient depletion of lung glutathione by diethylmaleate enhances oxygen toxicity. J. Appl. Physiol. 85, 571–574.

Dinarello, C.A., & Wolff, S.M. The role of interleukin-1 in disease. N. Engl. J. Med. 328, 106–113.

Fowler, A.A., Walchak, S., Giclas, P.C., Henson, P.M., & Hyers, T.M. (1982). Characterization of antiproteinase activity in the adult respiratory distress syndrome. Chest 81(Suppl), 50S–51S.

Fowler, A.A., Hamman, R.F., Good, J.T., Benson, K.N., Baird, M., Eberle, D.J., Petty, T.L., & Hyers, T.M. (1983). Adult respiratory distress syndrome: Risk with common predispositions. Ann. Intern. Med. 98, 593–597.

Frank, L., Summerville, J., & Massaro, D. (1980). Protection from oxygen toxicity with endotoxin: Role ofthe endogenous antioxidant enzymes of the lung. J. Clin. Invest. 65, 1104–1110.

Frank, L., Wood, D.L., & Roberts, R.J. (1981). Effect of diethyldithiocarbamate on oxygen toxicity and lung enzyme activity in immature and adult rats. Biochem. Pharmacol. 27, 51–254.

Freeman, B.A., Young, S.L., & Crapo, J.D. (1983). Liposome-mediated augmentation of superoxide dismutase in endothelial cells prevents oxygen injury. J. Biol. Chem. 258, 12534–12542.

Gadek, J.E. (1992). Adverse effects of neutrophils on the lung. Am. J. Med. 92(6A), 27S-31S.

Grum, C.M., Ragsdale, R.A., Ketai, L.H., & Simon, R.H. (1987). Plasma xanthine oxidase activity in patients with ARDS. J. Crit. Care. 2, 22–26.

Guidot, D.M., McCord, J.M., Wright, R.M., & Repine, J.E. (1993). Absence of electron transport (Rho° state) restores growth of a manganese-superoxide dismutase-deficient saccharomyces cerevisiae in hyperoxia. J. Biol. Chem. 268, 26699–26703.

Guidot, D.M., Stevens, E.E., Repine, M.J., Lucca-Broco, A., 7 Repine, J.E. (1994a). Intratracheal but not intravascular interleukin-1 causes acute edematous injury in isolated neutrophil-perfused rat lungs through an oxygen radical mediated mechanism. J. Lab. Clin. Med. 123, 605–609.

Guidot, D.M., Repine, M.J., Westcott, J.Y., & Repine, J.E. (1994b). Intrinsic 5-lipoxygenase activity is required for neutrophil responsivity. Proc. Natl. Acad. Sci. (USA) 91, 8156–8159.

Guidot, D.M., Repine, J.E., Kitlowski, A.D., Flores, S.C., Nelson, S.K., Wright, R.M., & McCord, J.M. (1995). Mitochondrial respiration scavenges extramitochondrial superoxide anion via a nonenzymatic mechanism. J. Clin. Invest. 96, 1131-1136.

Hammerschmidt, D.E., Weaver, L.J., Hudson, L.D., Craddock, P.R., & Jacob, H.S. (1980). Association of complement activation and elevated plasma-C5a with Adult Respiratory Distress Syndrome. Pathophysiological relevance and possible prognostic value. Lancet 3, 947–949.

Harada, R.N., Vatter, A.E., & Repine, J.E. (1984). Macrophage effector function in pulmonary oxygen toxicity: Hyperoxia damages and stimulates alveolar macrophages to make and release chemotaxins for polymorphonuclear leukocytes. J. Leukocyte Biol. 35, 373–383.

Heffner, J.E., & Repine, J.E. (1989). Pulmonary strategies of antioxidant defense. Am. Rev. Respir. Dis. 140, 531–554.

Heidel, J.R., Sassenfeld, H.M., Maliszewdki, C.R., Silflow, R.M., Baker, P.E., Taylor, S.M, & Leid RW. (1990). Functional studies of bovine alveolar neutrophils elicited with recombinant bovine IL-1 beta. J. Immunol. 144(3), 1037–1041.

Henson, P.M., Henson, J.E., Fitlschen, C., Kimani, G., Bratton, D.L., & Riches, D.W.H. (1988). Phagocytic cells: degranulation and secretion. In: Inflammation: Basic Principles and Clinical

Correlates. (Gallin, J.I., Goldstein, I.M., & Snyderman, R., Eds.), pp. 363–380. New York, Raven Press.

Hosea, S., Brown, E., Hammer, C., & Frank, M. (1980). Role of complement activation in model of the adult respiratory distres ssyndrome. J. Clin. Invest. 375–382.

Huang, S., Paulauskis, J.D., Godleski, J.J., & Kobzik, L. (1992). Expression of macrophage inflammatory protein-2 and KC mRNA in pulmonary inflammation. Am. J. Path. 141, 981–988.

Hybertson, B.M., Leff, J.A., Beehler, C.J., Barry, P.C., & Repine, J.E. (1995a). Effect of vitamin E deficiency and supercritical fluid aerosolized vitamin E supplementation on interleukin-1-induced oxidative lung injury in rat. Free Rad. Biol. Med. 18, 537–542.

Hybertson, B.M., Lampey, A.S., Clarke, J.H., Koh, Y., & Repine, J.E. (1995b). N-acetylcysteine pretreatment attenuates paraquat-induced lung leak in rats. Redox Report 1, 337-342.

Iqbal, J., Clerch, L.B., Hass, M.A., Frank, L., & Massaro, D. (1989). Endotoxin increases lung Cu,Zn superoxide dismutase mRNA: °2 raises enzyme synthesis. Am. J. Physiol. 257, L61–L64.

Jackson, R.M. & Frank, L. (1984). Ozone-induced tolerance to hyperoxia in rats. Am. Rev. Resp. Dis. 129, 425–429.

Jacobs, R.F., Tabor, D.R., Burks, A.W., & Campbell, G.D. (1989). Elevated interleukin-1 release by human alveolar macrophages during the adult respiratory distress syndrome. Am. Rev. Resp. Dis.140(6), 1686–1692.

Janoff, A. (1985). Elastase in tissue injury. Ann. Rev. Med. 36, 207–216.

Koh, Y., Hybertson, B.M., Jepson, E.K., Cho, O.J., & Repine, J.E. (1995a). Cytokine-induced neutrophil c hemoattractant is necessary for interleukin-1-induced lung leak in rats. J. Invest. Med. 43, 143A. (Abstract).

Koh, Y., Hybertson, B.M., Jepson, E.K., Cho, O.J., & Repine, J.E. (1995b). Cytokine-induced neutrophil chemoattractant is necessary for interleukin-1-induced lung leak in rats. J. Appl. Physiol. 79, 472-478.

Laufe, M.D., Simon, R.H., Flint, A., & Keller, J.B. (1986). Adult respiratory distress syndrome in neutropenic patients. Am. J. Med. 80, 1022–1026.

Leff, J.A., Oppegard, M.A., McCarty, E.C., Wilke, C.P., Shanley, P.F., Day, C.E., Ahmed, N.K., Patton, L.M., & Repine, J.E. (1992). Dimethyl sulfoxide decreases lung neutrophil sequestration and lung leak. J. Lab. Clin. Med. 120, 282–289.

Leff, J.A., & Repine, J.E. (1993). Neutrophil-mediated tissue injury. In: The Natural Immune System: The Neutrophil. (Abramson, J.S., & Wheeler, J.G., Eds.), pp. 229–262. Oxford: Oxford University Press.

Leff, J.A., Parsons, P.E., Day, C.E., Taniguchi, N., Jochum, M., Fritz, H., Moore, F.A., Moore, E.E., McCord, J.M., & Repine, J.E. (1993a). Serum antioxidants as predictors of the Adult Respiratory Distress Syndrome in septic patients. Lancet 341, 777–780.

Leff, J.A., Wilke, C.P., Hybertson, B.M., Shanley, P.F., Beehler, C.J., & Repine, J.E. (1993b). Post-insult treatment with N-acetylcysteine decreases interleukin-1-induced lung neutrophil sequestration and oxidative lung leak in rats. Am. J. Physiol. 265, L501–L506.

Leff, J.A., Baer, J.W., Kirkman, J.M., Bodman, M.E., Shanley, P.F., Cho, O.J., Ostro, M.J., & Repine, J.E. (1994c). Post-treatment with liposome-entrapped prostaglandin E_1 decreases interleukin-1α-induced neutrophil accumulation and lung leak in rats. J. Appl. Physiol. 76, 151–157.

Leff, J.A., Baer, J.W., Bodman, M.E., Kirkman, J.M., Shanley, P.F., Patton, L.M., Beehler, C.J., McCord, J.M., & Repine, J.E. (1994d). Interleukin-1-induced lung neutrophil accumulation and oxygen metabolite-mediated lung leak in rats. Am. J. Physiol. 266, 2–8.

Leff, J.A., Bodman, M.E., Cho, O.J., Rohrbach, S., Reiss, O.K., Vannice, J.L., & Repine. J.E. (1994e). Post-insult treatment with interleukin- 1 receptor antagonist decreases oxidative lung injury in rats given interleukin- 1 intratracheally. Am. J. Respir. Crit. Care Med. 105, 109–112.

Lemanske, R.F., Guthman, D.A., Oertel, H., Barr, L., & Kaliner, M. (1983). The biological activity of mast cell granules. VI. The effect of vinblastine-induced neutropenia on rat cutaneous late phase reactions. J. Immunol. 130(6), 2837–2842.

Maier, K.L., Mateikova, E., Hinze, H., Leuschel, L., Weber, H., & Beck-Speier, I. (19889). Different selectivities of oxidants during oxidation of methionine residues in the α-1-proteinase inhibitor. FEBS Lett. 250, 221–226.

Martin, W.J., Gadek, J.E., Hunninghake, G.W., & Crystal, R.G. (1981). Oxidant injury of lung parenchymal cells. J. Clin. Invest. 68, 1277–1288.

Martin, W.J. (1984). Neutrophils kill pulmonary endothelial cells by a hydrogen-peroxide-dependent pathway; and *in vitro* model of neutrophil-mediated lung injury. Am. Rev. Resp. Dis. 130, 209–213.

Martin, T.R., Pistorese, B.P., Hudson, L.D., & Maunder, R.J. (1991). The function of lung and blood neutrophils in patients with the adult respiratory distress syndrome. Am. Rev. Resp. Dis. 144, 254–262.

McCord, J.M., & Fridovich, I. (1969). Superoxide dismutase: an enzymic function for erythrocuprein (hemocuprein). J. Biol. Chem. 244, 6049–6055.

McCord, J.M. (1974). Free radicals and inflammation: Protection of synovial fluid by superoxide dismutase. Science 185, 529–531.

McCord, J.M., & Omar, B.A. (1993). Sources of free radicals. Toxicol. Industr. Health 9, 23–37.

McGuire, W.W., Spragg, R.G., Cohen, A.B., & Cochrane, C.G. (1982). Studies on the pathogenesis of the adult respiratory distress syndrome. J. Clin. Invest. 69, 543–553.

Miller, M.D., & Krangel, M.S. (1992). Biology and biochemistry of the chemokines: A family of chemotactic and inflammatory cytokines. Crit. Rev. Immunol. 12, 17–46.

Milligan, S.A., Hoeffel, J.M., Goldstein, I.M., & Flick, M.R. (1988). Effect of catalase on endotoxin-induced acute lung injury in unanesthetized sheep. Am. Rev. Resp. Dis. 137, 420–428.

Nelson, S.K., Bose, S.K., & McCord, J.M. (1994). The toxicity of high-dose superoxide dismutase suggests that superoxide can both initiate and terminate lipid peroxidation in the reperfused heart. Free. Rad. Biol. Med. 16, 195–200.

Ognibene, F.P., Martin, S.E., Parker, M.M., Schlesinger, T., Roach, P., Burch, C., Shelhamer, J.H., Parrillo, J.E. (1986). Adult respiratory distress syndrome in patients with severe neutropenia. N. Engl. J. Med. 315, 547–551.

Pacht, E.R., Timerman, A.P., Lykens, M.G., & Merola, A.J. (1991). Deficiency of alveolar fluid glutathione in patients with sepsis and the adult respiratory distress syndrome. Chest 100, 1397–1403

Padmanabhan, R.V., Gudapaty, R., Liener, I.E., Schwartz, B.A., & Hoidal, J.R. (1985). Protection against pulmonary oxygen toxicity in rats by intratracheal administration of liposome-encapsulated superoxide dismutase or catalase. Am. Rev. Respir. Dis. 132, 164–167.

Pahl, H.L., Banerle, P.A. Oxygen and the control of gene expression. Bioessays 1994;16:497.

Parrish, D.A., Mitchell, B.C., Henson, P.M., & Larsen, G.L. (1984). Pulmonary responses of fifth component of complement-sufficient and -deficient mice to hyperoxia. J. Clin. Invest. 74, 956–965.

Parsons, ,P.E., Worthen, G.S., Moore, E.E., Tate, R.M., & Henson, P.M. (1989). The association of circulating endotoxin with the development of the adult respiratory distress syndrome. Am. Rev. Resp. Dis. 140, 294–301.

Parsons, P.E., Worthen, G.S., Moore, F.A., Moore, E.E., & Henson, P.M. (1990). Activity levels of circulating neutrophils in acute lung injury. Am. Rev. Resp. Dis.141, A511. (Abstract)

Patterson, C.E., Butler, J.A., Byrne, F.D., & Rhodes, M.L. (1985). Oxidant lung injury: Intervention with sulfhydryl reagents. Lung 163, 23–32.

Patton, L.M., Saggart, B.S., Ahmed, N.K., Leff, J.A., & Repine, J.E. (1995). Interleukin-1β induced neutrophil recruitment and acute lung injury in hamsters. Inflammation 19, 23–29.

Petrone, W.F., English, D.K., Wong, K., & McCord, J.M. (1980). Free radicals and inflammation: Superoxide-dependent activation of a neutrophil chemotactic factor in plasma. Proc. Natl. Acad. Sci. USA 77, 1159–1163.

Pryor, W.A., Dooley, M.M., & Church, D.F. (1986). The mechanisms of the inactivation of human alpha-1-proteinase inhibitor by gas-phase cigarette smoke. Adv. Free Radical Biol. Med. 2, 161–188.

Repine, J.E., White, J.G., Clawson, C.C., & Holmes, B.M. (1974). Effect of phorbol myristate acetate on the metabolism and ultrastructure of neutrophils in chronic granulomatous disease. J. Clin. Invest. 54, 83–90.

Repine, J.E., Cheronis, J.C., Rodell, T.C., Linas, S.L., & Patt, A. (1987). Pulmonary oxygen toxicity and ischemia-reperfusion injury. A mechanism in common involving xanthine oxidase and neutrophils. Am. Rev. Respir. Dis. 136, 483–485.

Repine, J.E. (1992). Scientific perspectives on adult respiratory distress syndrome. Lancet 339, 466–469.

Repine, J.E. (1994). Interleukin-1-mediated acute lung injury and tolerance to oxidative injury. Environ. Health Perspect. 102 (Supplement), 75-78.

Rinaldo, J.E., & Borovetz, H. (1985). Deterioration of oxygenation and abnormal lung microvascular permeability during resolution of leukopenia in patients with diffuse lung injury. Am. Rev. Resp. Dis. 131, 579–583.

Rocker, G.M., Pearson, D., Wiseman, M.S., & Shale, D.J. (1989). Diagnostic criteria for adult respiratory distress syndrome: Time for reappraisal. Lancet 1, 120–123.

Sacks, T., Moldow, C.F., Craddock, P.R., Bowers, T.K., & Jacob, H.S. (1978). Oxygen radicals mediate endothelial cell damage by complement-stimulated granulocytes. J. Clin. Invest. 61, 1161–1167.

Schraufstatter, I.U., Hinshaw, D.B., Hyslop, P.A., Spragg, R.G., & Cochrane, C.G. (1986). Oxidant injury of cells. J. Clin. Invest. 77, 1312–1320.

Seeger, W., Grimminger, F., Barden, M., Becker, G., Lohmeyer, J., Heinrich, D., & Lasch, H.G. (1991). Omega-oxidized leukotriene B4 detected in the bronchoalveolar lavage fluid of patients with non-cardiogenic pulmonary edema, but not in those with cardiogenic edema. Int. Care Med.17, 1–6.

Shasby, D.M., Vanbenthuysen, K.M., Tate, R.M., Shasby, S.S., McMurtry, I.F., & Repine, J.E. (1982a). Granulocytes mediate acute edematous lung injury in rabbits and isolated rabbit lungs perfused with phorbol myristate acetate: Role of oxygen radicals. Am. Rev. Resp. Dis. 125, 443–447.

Shasby, D.M., Fox, R.B., Harada, R.N., & Repine, J.E. (1982b). Reduction of the edema of acute hyperoxic lung injury by granulocyte depletion. J. Appl. Physiol. 52, 1237–1244.

Shasby, D.M., Shasby, S.S., & Peach, M.J. (1983). Granulocytes and phorbol myristate acetate increase permeability to albumin of cultured endothelial monolayers and isolated perfused lungs. Role of oxygen radicals and granulocyte adherence. Am. Rev. Respir. Dis. 127, 72–76.

Sies, H. (1987). Antioxidant activity in cells and organs. Am. Rev. Respir. Dis. 136, 478–480.

Siler, T.M., Swierkosz, J.E., Hyers, T.M., Fowler, A.A., & Webster, R.O. (1989). Immunoreactive interleukin-1 in bronchoalveolar lavage fluid of high-risk patients and patients with the adult respiratory distress syndrome. Exp. Lung Res. 15(6), 881–894.

Simon, R.H., Scoggin, C.H., & Patterson, D. (1981). Hydrogen peroxide causes the fatal injury to human fibroblasts exposed to oxygen radicals. J. Biol. Chem. 256, 7181–7186.

Southorn, P.A., & Powis, G. (1988). Free radicals in medicine. Mayo Clin. Proc. 63, 390–408.

Stephenson, A.H., Lonigro, A.J., Hyers, T.M., Webster, R.O., & Fowler, A.A. (1988). Increased concentrations of leukotrienes in bronchoalveolar lavage fluid of patients with ARDS or at risk for ARDS. Am. Rev. Resp. Dis.138, 714–719.

Suter, P.M., Suter, S., Girardin, E., Roux-Lombard, P., Grau, G.E., & Dayer, J. (1992). High bronchoalveolar levels of tumor necrosis factor and its inhibitors, interleukin- 1, interferon, and elastase, in patients with adult respiratory distress syndrome after trauma, shock, or sepsis. Am. Rev. Resp. Dis. 145, 1016–1022.

Sznajder, J.I., Fraiman, A., Hall, J.B., Sanders, W., Schmidt, G., Crawford, G., Nahum, A., Factor, P., & Wood, L.D.H. (1989). Increased hydrogen peroxide in the expired breath of patients with acute hypoxemic respiratory failure. Chest 96, 606–612.

Tate, R.M., & Repine, J.E. (1983a). Neutrophils and the adult respiratory distress syndrome. Am. Rev. Respir. Dis. 128, 552–559.

Tate, R.M., Morris, H.G., Schroeder, W.R., & Repine, J.E. (1984b). Oxygen metabolites stimulate thromboxane production and vasoconstriction in isolated saline-perfused rabbit lungs. J. Clin. Invest. 74, 608–613.

Tate, R.M., Morris, H.G., Schroeder, W.R., & Repine, J.E. (1984). Oxygen metabolites stimulate thromboxane production and vasoconstriction in isolated saline-perfused rabbit lungs. J. Clin. Invest. 74, 608–613.

Terada, L.S., Dormish, J.J., Shanley, P.F., Leff, J.A., Anderson, B.O., & Repine, J.E. (1992). Circulating xanthine oxidase mediates lung neutrophil sequestration after intestinal ischemia-reperfusion. Am. J. Physiol. 263, L394–L401.

Thet, L.A., Parra, S.C., & Shelburne, J.D. (1984). Repair of oxygen-induced lung injury in adult rats. The role of ornithine decarboxylase and polyamines. Am. Rev. Respir. Dis. 129, 174–181.

Till, G.O., Johnson, K.J., Kunkel, R., & Ward, P.A. (1982). Intravascular activation of complement and acute lung injury. Dependency on neutrophils and toxic oxygen metabolites. J. Clin. Invest. 69, 1126–1135.

Travis, J., & Salveson, G.S. (1983). Human plasma proteinase inhibitors. Ann. Rev. Biochem. 52, 655–709.

Turrens, J.F., Freeman, B.A., & Crapo, J.D. (1982). Hyperoxia increases hydrogen peroxide formation by lung mitochondria and microsomes. Arch. Biochem. Biophys. 217, 411–419.

Ulich, T.R., Yin, S.M., Guo, K.Z., del Castillo, J., Eisenberg, S.P., & Thompson, R.C. (1991). The intratracheal instillation of cytokines. III. The interleukin- 1 receptor antagonist inhibits endotoxin and IL-1-induced acute inflammation. Am. J. Pathol. 138, 521–524.

van Asbeck, B.S., Hoidal, J., Vercellotti, G.M., Schwartz, B.A., Moldow, C.F., & Jacob, H.S. (1985). Protection against lethal hyperoxia by tracheal insufflation of erythrocytes: role of red cell glutathione. Science 227, 756–759.

Van Kuijk, F.J.G.M., Sevanian, A., Handelman, G.J., & Dratz, E.A. (1987). A new role for phospholipaseA2: protection of membranes from lipid peroxidation damage. Trends Biochem. Sci. 12, 31–34.

Waintrub, M.L., Terada, L.S., Beehler, C.J., Anderson, B.O., Leff, J.A., & Repine, J.E. (1990). Xanthine oxidase is increased and contributes to paraquat-induced acute lung injury. J. Appl. Physiol. 68, 1755–1757.

Weiss, S.J. (1980). The role of superoxide in the destruction of erythrocyte targets by human neutrophils. J. Biol. Chem. 255, 9912–9917

Weiss, S.J., Young, J., LoBuglio, A.F., Slivka, A., & Nimeh, N.F. (1981). Role of hydrogen peroxide in neutrophil-mediated destruction of cultured endothelial cells. J. Clin. Invest. 68, 714–721.

White, C.W., Jackson, J.H., McMurtry, I.F., & Repine, J.E. (1988). Hypoxia increases glutathione redox cycle and protects rat lungs against oxidants. J. Appl. Physiol. 65, 2607–2616.

Wilson, W.C., Swetland, J.F., Benumof, J.L., Laborde, P., & Taylor, R. (1992). General anesthesia and exhaled breath hydrogen peroxide. Anesthesiology 76, 703–710.

Yoo, J.H., Erzurum, S.C., Hay, J.G., Lemarchand, P., & Crystal, R.G. (1994). Vulnerability of the human airway epithelium to hyperoxia. Constitutive expression of the catalase gene in human bronchial epithelial cells despite oxidant stress. J. Clin. Invest. 93, 297–302.

Yusa, T., Crapo, J.D., & Freeman, B.A. (1984). Hyperoxia enhances lung and liver nuclear superoxide generation. Biochim. Biophys. Acta 798, 167–174.

Zheutlin, L.M., Thonor, E.J., Jacobs, E.R., Hanley, M.E., Balk, R.A, & Bone, R.C. (1986). Plasma elastase levels in the adult respiratory distress syndrome. J. Crit. Care 1, 39–44.

Zimmerman, G.A., Renzetti, A.D., & Hill, H.R. (1983). Functional and metabolic activity of granulocytes from patients with adult respiratory distress syndrome. Am. Rev. Resp. Dis. 127, 290–300.

Zweier, J.L., Kuppusamy, P., & Lutty, G.A. (1988). Measurement of endothelial cell free radical generation: Evidence for a central mechanism of free radical injury in postischemic tissues. Proc. Natl. Acad. Sci. (USA) 85, 4046–4050.

NITRIC OXIDE REGULATION OF SUPEROXIDE AND PEROXYNITRITE-DEPENDENT REACTIONS

Homero Rubbo and Bruce A. Freeman

Advances in Molecular and Cellular Biology
Volume 25, pages 135-155.

ISBN: 0-7623-0379-4

I. NITRIC OXIDE—A FREE RADICAL SIGNAL TRANSDUCING AGENT

Nitric oxide (·NO, nitrogen monoxide) is an endogenously-synthesized free radical first characterized as a non-eicosanoid component of endothelial-derived relaxation factor (EDRF) (Furchgott, 1980; Palmer, 1987). Nitric oxide is produced by a variety of mammalian cells including endothelium, neurons, smooth muscle cells, macrophages, neutrophils, platelets, and type 2 pneumocytes (Moncada, 1991a). The biological actions of ·NO range from mediating vasodilation, neurotransmission, inhibition of platelet adherence/aggregation and the macrophage and neutrophil-mediated killing of pathogens. Nitric oxide is synthesized by the oxidation of L-arginine to L-citrulline in the presence of molecular oxygen and NADPH. The family of enzymes responsible for the synthesis of ·NO, nitric oxide synthases (NOS), are homologous to cytochrome P450 reductase, with both constitutive (cNOS) and inducible (iNOS) forms of NOS reported (Bredt, 1991; Mardsen, 1992). The iNOS is induced by different cytokines and by bacterial and parasitic antigens, which often result in the expression of cytokines as well.

Nitric oxide has proven to be a ubiquitous signal transduction molecule and a potent mediator of tissue injury because of its low molecular mass, volatility, lipophilicity, free radical nature, and diverse reactivities. Because of the transient nature of free radical species and their often broad range of reactivities, it becomes challenging to define the mechanisms of tissue injury in processes of oxidant stress when a diverse spectrum of reactive species is produced. Due to its chemical nature, ·NO is an active participant in tissue free radical reactions, readily reacting with partially-reduced oxygen species (i.e., $O_2^{\bullet -}$ and organic-derived free radicals). Nitric oxide is also a relatively less reactive iron ligand, forming paramagnetic species after reaction of ·NO with ferrous ions containing one or more additional coordinating anionic ligands (Kanner, 1991). Except for its rapid radical-radical reactions, nitric oxide has a relatively low reactivity for a free radical species, resulting in a long biological half-life ($t_{1/2}$ = 5-30 s, Knowles, 1993), however others have reported the half-life of ·NO *in vivo* to be considerably shorter (~0.1 s, Kelm, 1990). At low ·NO and oxygen concentrations, the half-life of ·NO will be considerably longer, consequently ·NO diffusion distances will be much greater (Vanderkooi, 1994). Nitric oxide and oxygen closely resemble each other in diffusivity, with ·NO slightly less lipophilic, resulting in a greater apparent rate of ·NO diffusion in protein and lower apparent rate of ·NO diffusivity in lipid, both relative to oxygen (Vanderkooi, 1994).

The chemical properties of ·NO which make it a useful signal transduction agent include (a) relatively low reactivity for being a free radical species, resulting in a biological half-life in the range of seconds, (b) charge neutrality, a small molecular radius and hydrophobicity, allowing facile transmembrane diffusion, (c) its selective reactivity with heme, iron-sulfur and thiol-containing proteins which participate in signal transduction mechanisms, and (d) its ability to react with molecular oxygen and oxygen-derived free radicals, thus providing tissues with a non-enzymatic method for modulating the local concentration of ·NO. Another important feature for the biologic transport and delivery of nitric oxide may also be its ability to readily form nitroso-thiol complexes (i.e., S-nitroso-albumin or S-nitroso-glutathione, Stamler, 1992a). More recently Scharfstein et al. (1994) reported that transnitrosation between plasma protein thiol-bound ·NO and low molecular weight thiol pools occurs *in vivo.* From this, it can be speculated that S-nitroso-albumin may serve in part as an extracellular reservoir of ·NO which is transferred more efficiently to the intracellular milieu by smaller, less diffusion-limited S-nitrosothiol species or membrane-associated thiol-containing proteins.

II. NITRIC OXIDE REACTIONS WITH OXYGEN AND SUPEROXIDE

Nitric oxide will react with molecular oxygen in aqueous solution, yielding nitrite (NO_2^-) and nitrate (NO_3^-) as decomposition products of the intermediates N_2O_3 and N_2O_4 (Kharitonov, 1994). This reaction is second order for + ·NO, with tissue ·NO levels in the range of 1-1000 nM, and first order for O_2 (k= 6-8 x 10^6 M^{-2} s^{-1}), thus this reaction does not account for the long $t_{1/2}$ of ·NO or serve as a principal pathway for ·NO reaction *in vivo*. A critical reaction that ·NO undergoes in oxygenated biologic media is a direct bimolecular reaction with superoxide anion ($O_2^{\bullet -}$).

$$O_2^{\bullet -} + \cdot NO \rightarrow {}^-OONO \tag{1}$$

$$H+ + {}^-OONO \rightarrow HOONO \qquad (pKa=6.8) \tag{2}$$

The rapid reaction of ·NO with $O_2^{\bullet -}$ yields peroxynitrite ($ONOO^-$) at almost diffusion-limited rates (6.7 x 10^9 m^{-1} s^{-1}, Huie, 1993). This rate constant is 3.5 times faster than the enzymatic disproportionation of $O_2^{\bullet -}$ catalyzed by superoxide dismutases at neutral pH ($k_{SOD} = 2 \times 10^9$ m^{-1} s^{-1}). Thus, $ONOO^-$ formation represents a major potential pathway of ·NO reactivity that depends on both rates of tissue ·NO and $O_2^{\bullet -}$ production, with steady state $O_2^{\bullet -}$ concentrations ranging from 10 pM under basal conditions to 0.01 to 0.1 μM during tissue pathologic states (Zweier, 1989). Peroxynitrite has a half-life of <1s under physiological conditions, due to proton-catalyzed decomposition and target molecule reactions (Beckman, 1990; Radi, 1991a; Ohara, 1994). Thus, the reaction of ·NO with $O_2^{\bullet -}$, initially viewed as a route for ·NO "inactivation," instead yields the potent oxidizing species $ONOO^-$, which exhibits direct oxidative reactivities and will protonate to peroxynitrous acid at biological pH (ONOOH, pK_a=6.8).

Compared with $ONOO^-$, ONOOH has its own unique reactivities, many of which may possibly account for reactions previously ascribed to the reactive species formed following transition metal-catalyzed reduction of hydroperoxides. It was initially observed that ONOOH had a hydroxyl radical (·OH)-like reactivity and/or reacted as an activated species in a manner akin to ·OH, based on product analyses and the action of OH scavengers (Beckman, 1990). Thermodynamic calculations do not encourage that ONOOH will physically separate to yield free ·OH and nitrogen dioxide (·NO_2, Koppenol, 1992) as originally proposed by Mahoney (1970) and more recently reviewed in detail by Edwards and Plumb (1993). Mechanistic explanations which have been advanced to account for this intriguing and controversial reactivity of ONOOH are based on the general concept that ONOOH reacts as an excited species. This activated state of ONOOH, for expediency, can be expressed as ONOOH*. One explanation which could explain the unique and potent ·OH-like reactivity of ONOOH* is the formation of a vibrationally-excited intermediate species in the isomerization of HOON = O to $HONO_2$. This species could represent ONOOH in a distorted trans configuration having a lengthened O-O bond and the terminal oxygen near the nitrogen (Koppenol, 1992). The bottom line should not be lost in this controversy, however, if ONOOH* looks and smells like a skunk (i.e., ·OH), then for envisioning many of the biological sequelae of ONOOH* reactivity, it may as well be so. It also then becomes a challenging exercise in biochemistry to ferret the underlying mechanism(s) of critical events occurring during processes of oxidant tissue injury including hydroxylation of aromatic amino acids (van der Vliet, 1994a) or DNA bases (King, 1994), initiation of membrane or lipoprotein lipid peroxidation and oxidation of the sulfhydryl and iron-sulfur complexes so critical to intermediary metabolism. For these reasons, it becomes important to consider the critical influences that ·NO will exert on oxygen radical-dependent reactions. This is especially the case in pathologic events which exhibit a coordinated stimulation of oxygen radical production and enhanced tissue rates of ·NO production.

III. TARGET MOLECULE REACTIONS OF ·NO AND $ONOO^-$

Nitric oxide has been recognized as a key macrophage-derived effector molecule, defending the host against bacteria, tumor cells and parasites (Table 1). Also, excess tissue ·NO production may lead to the pathological responses occurring during such diverse events as allograft transplant rejection, tissue ischemia-reperfusion phenomena, neuro-excitotoxin-induced brain injury and immune complex-stimulated pulmonary edema (Matheis, 1992; Snyder, 1993; Mulligan, 1991, 1992). Metal-and thiol-containing proteins serve as major target sites for ·NO reactions. The toxicity of ·NO has principally been attributed to direct ·NO reaction with thiol and iron-sulfur containing mitochondrial enzymes (Hibbs, 1988) and the inhibition of DNA synthesis via inactivation of the non-heme iron enzyme ribonucleotide reductase (Lepoivre, 1994). Nitric oxide also mediates inhibition of the mitochondrial cytochrome c oxidase (Cleeter, 1994) and deenergizes mitochondria at low ·NO and oxygen concentrations (Schweizer, 1994). Heme proteins such as

Table 1. Enzyme and Protein Targets of Nitric Oxide and Peroxynitrite

Reaction site	*Nitric oxide*	*Peroxynitrite*
Heme-iron	guanylate cyclase	
	cytochrome c oxidase	
Non-heme iron /metal	cyclooxygenase*	mitochondrial aconitase
	lipoxygenase	NADH dehydrogenase
	cytosolic aconitase	cytochrome c oxidase
	mitochondrial aconitase*	ATP ase CuZn and Mn
	xanthine oxidase*	superoxide dismutase
	cytochrome P-450	
	hemoglobin	
	mitochondrial respiratory complex I	
	mitochondrial respiratory complex II	
	ribonucleotide reductase	
Thiols	GAPDH	GAPDH
	PTP ase	succinate dehydrogenase
	protein kinase C	succinate dehydrogenase#
	adenyl cyclase	fumarate reductase#
	G proteins	
	K^+ and Ca^{+2} channels	
	NADPH oxidase	

Abbreviations: mitochondrial complex I and II, NADH and succinate ubiquinone oxidoreductases, respectively; GAPDH, glyceraldehyde-3-phosphate dehydrogenase; PTP ase, low Mr phosphotyrosine protein phosphatase; # represents energetic enzymes studied in *Trypanosoma cruzi* , and * represents mammalian enzymes inactivated by high (and not biologically relevant) concentrations of nitric oxide.

guanylate cyclase are activated by low ·NO concentrations (Stamler, 1992b), while cytochrome P-450 (Wink, 1993a) and other nonheme iron proteins such as aconitase (Hibbs, 1988), lipoxygenase (Kanner, 1992) and xanthine oxidase (Fukahori, 1994) are inhibited by ·NO, albeit minimally or from much greater concentrations of ·NO than biologically achievable. Since tissue ·NO concentrations are low ($<1 \mu M$), it is important to note that significant reactivity with heme, iron-sulfur and thiol-containing proteins often require high concentrations of ·NO or ·NO-generating agents. For example, mitochondrial aconitase has been indicated as a major target of ·NO-mediated toxicity in several cell types. Similarly, ·NO has been implicated in the activation of iron-responsive element-binding protein in macrophages (cytosolic aconitase, Drapier, 1993). Recently, it has been reported that anaerobic exposure of isolated aconitase to high concentrations of ·NO ($> 100 \mu M$) led to the formation of a reversible inhibitory complex between ·NO and the active site of the enzyme (Castro, 1994).

Thiol-containing enzymes are also critical targets for ·NO. The active site thiol of glyceraldehyde-3-phosphate dehydrogenase (GAPDH) is subject to ·NO, ·NO donor and $ONOO^-$-dependent reactions. This in turn leads to reaction with NAD^+, thus initiating non-enzymatic ADP-ribosylation reactions (Mohr, 1994). S-nitrosylation of the active site thiol of GAPDH is a prerequisite for subsequent

NAD^+ reaction, emphasizing the role of ·NO in the initial step of this pathway, long indicative of oxidant tissue injury. Nitric oxide also causes the inactivation of two members of the low M_r phosphotyrosine protein phosphatase family, sulfhydryl containing enzymes involved in a number of biological processes such as the transmission of mitotic signals at different levels of the mitotic activation cascade (Caselli, 1994).

Nitric oxide can conceivably potentiate many aspects of $O_2^{\bullet -}$-mediated tissue damage via $ONOO^-$ formation. To date, it has been shown that $ONOO^-$ is a potent oxidant capable of (a) directly oxidizing protein and non-protein sulfhydryls (Radi, 1991a; Rubbo, 1994a), (b) protonating to ONOOH, which exhibits unique and ·OH- like reactions via metal-independent mechanisms (Beckman, 1990; Radi, 1991a, 1991b; Ohara, 1994), and (c) reaction with metal centers to yield a species with the reactivity of nitronium cation (NO_2^+), an oxidizing and nitrating intermediate (Ischiropoulos, 1992a). In many instances, it is also becoming apparent that $ONOO^-$ can serve as a mediator in oxidative actions originally attributed to ·NO or other oxygen-derived species. Nitric oxide will potentiate $O_2^{\bullet -}$-mediated tissue damage and leads to $ONOO^-$ formation, which represents a major potential pathway of ·NO reactivity. Peroxynitrite is now being revealed to be a key contributing reactive species in pathological events associated with stimulation of tissue production of ·NO, i.e., systemic hypotension, inhibition of intermediary metabolism, ischemia-reperfusion injury, immune complex-stimulated pulmonary edema, cytokine-induced oxidant lung injury, and inflammatory cell-mediated pathogen killing/host injury (Zhu, 1992; Denicola, 1993; Carreras, 1994; Radi, 1994; Castro, 1994; Haddad, 1993; Matheis, 1992; Beckman, 1991; Mulligan, 1991, 1992). There is growing evidence that ·NO-mediated production of $ONOO^-$ readily occurs in vivo, underscoring the importance of understanding the target molecule reactions occurring during the coordinated production of oxygen and nitrogen-containing reactive species (Beckman, 1994; White, 1994; Carreras, 1994; Kooy, 1994).

IV. PRO-OXIDANT REACTIONS OF $ONOO^-$

- Peroxynitrite is 2000-fold more potent than hydrogen peroxide (H_2O_2) in oxidizing protein and non-protein sulfhydryl groups, with this process mediated by both one-and two-electron transfer reactions. Low yields of S-nitrosylated thiol products are also generated following $ONOO^-$ reaction with sulfhydryl compounds (Radi, 1991a; Pryor, 1994; Wu, 1994; Moro, 1994).
- $ONOO^-$ readily catalyzes membrane lipid peroxidation by metal-independent mechanisms (Radi, 1991b).
- $ONOO^-$ oxidizes DNA bases following protonation to ONOOH (Beckman, 1990; King, 1994).

- $ONOO^-$ is a major macrophage-derived oxidant, produced at rates of up to 0.5-1 $mM.min^{-1}.10^6$ cells (Ischiropoulos, 1992b). This observation, based in rodent cells, has recently been confirmed to occur in human neutrophils as well (Carreras, 1994).
- $ONOO^-$ is a bactericidal and trypanocidal agent produced by activated macrophages, killing *Escherichia coli* or *Trypanosoma cruzi* with a LD50 of 250 and 150 μM respectively (Zhu, 1992; Denicola, 1993).
- $ONOO^-$ exposure of rat heart mitochondria results in significant inactivation of electron transport components such as succinate dehydrogenase, NADH dehydrogenase and the mitochondrial ATPase (Radi, 1994).
- $ONOO^-$ reacts with iron-sulfur centers of mitochondrial respiratory chain proteins with a greater rate constant than for either $O_2^{\bullet -}$ or ·NO alone (Castro, 1994).
- $ONOO^-$ inactivates thiol-containing enzymes in mammals and parasitic cells, such as glyceraldehyde-3-phosphate dehydrogenase, succinate dehydrogenase and fumarate reductase (Mohr, 1994; Rubbo, 1994a).
- $ONOO^-$ formation and toxicity occurs during and accounts for much of the oxidant stress observed in porcine myocardial ischemia-reperfusion injury (Matheis, 1992).
- $ONOO^-$ causes depletion of key plasma antioxidants and damages proteins and lipids of human plasma (van der Vliet, 1994b).
- $ONOO^-$ modifies human low density lipoprotein (LDL) to a potentially atherogenic form recognized by the macrophage scavenger receptor, thus increasing the susceptibility of LDL to oxidation by transition metal ions and by depletion of vitamin E content (Hogg, 1993a; Graham, 1993).
- $ONOO^-$ reacts with metal chelates, as well as the metal centers of Cu,Zn-, Mn-, and Fe-superoxide dismutases to generate the potent nitrating agent NO_2^+, capable of nitrating tyrosine residues (Beckman, 1990; Ischiropoulos, 1992a).
- $ONOO^-$ is involved in extensive nitration of tyrosine in vascular endothelium, macrophage-derived foam cells and lesions associated with atheroma in human coronary arteries (Beckman, 1994; White, 1994). This is compelling evidence for $ONOO^-$ production and participation in oxidative processes associated with atherogenesis.
- $ONOO^-$ is involved in extensive nitration of tyrosine in septic and oxidant-stressed rodent and human lung tissue (Haddad, 1994).
- $ONOO^-$ decreases pulmonary alveolar epithelial type II cell sodium transport by damaging apically-located amiloride-sensitive sodium channels, and mediates inhibition of pulmonary surfactant function by lipid peroxidation and oxidation of surfactant proteins (Hu, 1994; Haddad, 1993).
- $ONOO^-$ has paradoxical effects on platelets, showing both pro- and antiaggregatory actions, which depend on the environment in which this oxidant is generated (Moro, 1994).

- $ONOO^-$ induces a concentration-dependent vasodilatation of the rat coronary vasculature probably due to the local formation of S-nitrosothiols (Villa, 1994). In contrast, impaired vascular reactivity results from the production of $ONOO^-$ in a cholesterol-fed rabbit model of hypercholesterolemia/atherogenesis (White, 1994).

V. ANTIOXIDANT REACTIONS OF ·NO

Despite the compelling evidence that the cogeneration of $O_2^{\bullet -}$ and ·NO often leads to cytotoxic events *via* direct toxic reactions, as well as from the generation of secondary products such as $ONOO^-$, there are a growing number of observations indicating that ·NO can sometimes serve a protective role in pathologic events associated with excess production and reaction of partially reduced oxygen species. One example of what was initially a suprising event to us was the observation that at low concentrations (similar to those produced *in vivo*) ·NO has been administered safely (to our knowledge) in the gas phase for weeks when treating pulmonary hypertension (Rossaint, 1994).

Numerous studies of cell or metal-induced lipoprotein oxidation reactions, as well as hepatic, cerebrovascular, splancnic, pulmonary and myocardial inflammatory and ischemia-reperfusion injury studies have recently shown that stimulation of endogenous ·NO production or exogenous administration of ·NO often inhibits oxidant-related mechanisms and blunts the ultimate expression of tissue injury at both molecular and functional levels (Guo, 1994; Yao, 1992; Clancy, 1992; Siegfried, 1992; Lefer, 1993; Wink, 1993b; Hogg, 1993b; Malo-Ranta, 1994; Jessup, 1993; Payne, 1993; Kubes, 1991, 1993; Kucrose, 1994; Choi, 1993; Kavanagh, 1994). In fact, dietary supplementation of L-arginine has recently been reported to limit the intimal thickening and loss of endothelial (i.e., ·NO)-dependent relaxation in the coronary artery of hypercholesterolemic rabbits and humans (Drexler, 1991, 1994).

The protective effects observed for ·NO with in vivo models of reperfusion injury, when administered as a bolus of nitrosothiol or other ·NO donors, are often ascribed to ·NO inhibition of inflammatory cell margination and function (Kubes, 1991; Siegfried, 1992; Lefer, 1993; Kurose, 1994). In addition, there is a rapid fall in ·NO levels upon reperfusion, thus augmenting ·NO-mediated signal transduction pathways at the level of cGMP provides a useful pharmacological approach for normalizing vascular function and blood flow in the critical early stages following tissue ischemia (Pinsky, 1994). In some of these models, inhibition of ·NO synthesis enhanced injury, again emphasizing the salutary role sometimes observed for ·NO in oxidant injury-related processes.

There are also other biochemical reactivities of ·NO that can account for its tissue-protective actions in events typically associated with oxidant stress. Nitric oxide not only stimulates $O_2^{\bullet -}$ induced lipid and lipoprotein oxidation and mediates oxidant-protective reactions in membranes, but will inhibit $O_2^{\bullet -}$ and $ONOO^-$-induced lipid oxidation at higher rates of ·NO production. The prooxidant versus antioxidant outcome of these reactions that are sensitive to ·NO regulation are critically dependent on relative concentrations of individual reactive species (Rubbo, 1994b, 1995). For example, the continuous infusion of various concentrations of ·NO (0-3 μM.min^{-1}) into liposome suspensions exposed to xanthine oxidase, acetaldehyde and EDTA-Fe^{+3} (1 μM.min^{-1} $O_2^{\bullet -}$ production) first stimulated and then inhibited formation of 2-thiobarbituric acid (TBA)-reactive products at greater rates of ·NO infusion (Figure 1, Table 2).

Peroxynitrite-dependent lipid peroxidation occurs by a predominantly iron-independent mechanism. Thus, TBA-reactive material formation will only be minimally inhibited by addition of diethylenetriaminepentaacetic acid (DTPA) to reactions, presumably via attenuation of $ONOO^-$ induced propagation reactions. Nitric oxide stimulates xanthine oxidase-dependent lipid peroxidation when production rates of ·NO are less than or equivalent to rates of $O_2^{\bullet -}$ production by xanthine oxidase. When the rate of ·NO production exceeds that of $O_2^{\bullet -}$, lipid peroxidation was inhibited. Since the cogeneration of $O_2^{\bullet -}$ and ·NO leads to the formation of $ONOO^-$, which readily induces oxidative damage to protein, lipid, carbohydrate, DNA, subcellular organelles and cell systems, the molecular mechanisms accounting for the

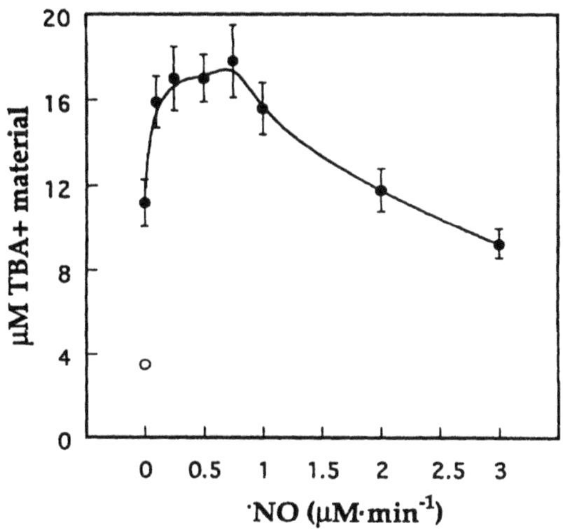

Figure 1. The influence of ·NO infusion on xanthine oxidase-dependent liposome oxidation. PC liposomes (6.6 mg.mL^{-1}) were oxidized with 1.5 mU.mL^{-1} XO, 10 mM acetaldehyde and 100 μM EDTA-Fe^{+3}. Infusion of ·NO (0.1, 0.25, 0.5, 0.7, 1, 2, and 3 μM.m^{-1}) was performed simultaneously. (o) represents incubation of reaction mixture in the absence of XO. Data represents the mean of duplicate determinations and was representative of three separate experiments. Adapted from (Rubbo, 1994) with permission.

Table 2. Nitric Oxide Inhibition of Xanthine Oxidase and Peroxynitrite-Dependent Liposome and Linolenic Acid Oxidation

Condition	*μM TBA+ material*
Control PC liposomes, EDTA-Fe^{+3}	1.9 ± 0.3
+ hypoxanthine	2.1 ± 0.3
+ hypoxanthine, XO	10.2 ± 0.7
+ hypoxanthine, XO, SNAP	3.8 ± 0.4
Control linolenic acid, EDTA-Fe^{+3}	1.1 ± 0.3
+ hypoxanthine, XO	9.6 ± 0.5
+ hypoxanthine, XO, SNAP	3.3 ± 0.4
Control PC liposomes	0.1 ± 0.03
+ $ONOO^-$	10.8 ± 1.3
+ EDTA-Fe^{+3}, $ONOO^-$	12.5 ± 1.8
+ EDTA-Fe^{+3}, $ONOO^-$, .NO	5.4 ± 0.9
+ EDTA-Fe^{+3}, $ONOO^-$, GSNO	5.8 ± 0.9
+ DTPA, $ONOO^-$	8.1 ± 0.6
+ DTPA, $ONOO^-$, .NO	3.4 ± 0.5
+ DTPA, $ONOO^-$, GSNO	3.9 ± 0.6

Notes: PC liposomes (6.6 $mg.mL^{-1}$) or linolenic acid (1 $mg{\cdot}mL^{-1}$) were oxidized for 1 hr instirred 100 mM potassium phosphate, pH 7.4, at 25°C (controls), using 150 mM hypoxanthine plus 10 $mU{\cdot}mL^{-1}$ XO or $ONOO^-$ (33 $\mu M.min^{-1}$ infusion for 15 m). The following additions were made as indicated: DTPA, 100 mM; EDTA-Fe^{+3}, 100 μM; S-nitrosoglutathione, 1 mM; S-nitrosopenicillamine, 100 μM and ·NO, via a 2.7 $\mu M{\cdot}m^{-1}$ infusion of ·NO gas dissolved in anaerobic buffer for 15 m. Data represents mean ± SD, n = 3. Adapted from (Rubbo, 1994) with permission.

prevention of reactive oxygen species-mediated injury by ·NO represent an important event, as well as a challenging experimental problem to define.

Several factors may account for the "antioxidant" role of ·NO:

A. Nitric Oxide Reaction with Iron

Because ·NO can serve as an iron ligand to form iron-nitrosyl complexes, it has been proposed that ·NO will modulate the pro-oxidant effects of iron and other transition metals, thereby limiting their role in the Haber-Weiss-catalyzed formation of ·OH and iron-dependent electron transfer reactions (Kanner, 1991). Treating Fenton reagents (such as Fe^{+2}-EDTA or myoglobin) with ·NO can inhibit H_2O_2-dependent ·OH production by forming nitrosyl-iron ligands. Similarly, ·NO will react with ferrylmyoglobin (X-[Fe(IV)=O] or HX-[Fe(IX)=O], the respective products of oxymyoglobin or myoglobin reaction with H_2O_2) to form iron- nitrosyl complexes which are less potent stimuli of LDL lipid oxidation. However, ESR spectroscopy analysis showed that the aerobic $O_2^{\bullet -}$-containing reaction systems (linoleic acid plus xanthine oxidase), which were exposed to lower and sometimes more biological rates of production of ·NO, did not yield detectable iron-nitrosyl complexes (Rubbo, 1994b,1995). This was not unexpected, because the rate of ·NO reaction with ferrous iron (2 × 10^7 M^{-1} s^{-1}, Traylor, 1992) is significantly slower than for ·NO reaction with either $O_2^{\bullet -}$ or lipid alkoxyl (LO·) and peroxyl (LOO·) radical species (Huie, 1993; Padmaja, 1993). It should be noted that with transition metals, ·NO can exert prooxidant effects as well, by reducing ferric iron complexes (Reif, 1990). This can induce the release of bound iron and indirectly substitute for other reductants in the Haber-Weiss reaction-mediated production of ·OH from H_2O_2.

B. Nitric Oxide Reactions with Alkoxyl and Peroxyl Radicals

Nitric oxide can act as an inhibitor of radical chain propagation reactions via radical-radical reaction with cytotoxic species such as lipid peroxyl or alkoxyl radicals. Under conditions where linolenic acid emulsions oxidized by xanthine oxidase or $ONOO^-$-derived reactive species showed inhibition of lipid peroxidation by ·NO (Table 2), mass spectral analysis of oxidation products showed formation of novel nitrogen-containing lipid adducts (Figures. 2a, 2b and 3) which were not detectable in the absence of ·NO sources. These novel lipid oxidation adducts are in part organic peroxynitrites before rearrangement and would be expected to occur in vivo when diverse inflammatory and pathological processes give rise to similar combinations of reactive species. Nitric oxide reacts with LO· and LOO· at near diffusion-limited rates (for LOO·, $k = 1.3 \times 10^9\ M^{-1} \cdot s^{-1}$, Padmaja, 1993). This rate constant is more than 2×10^3 times greater than for α-tocopherol reaction with peroxyl radicals ($k = 2.5 \times 10^6\ M^{-1} \cdot s^{-1}$), with tissue concentrations and lipophilicity of

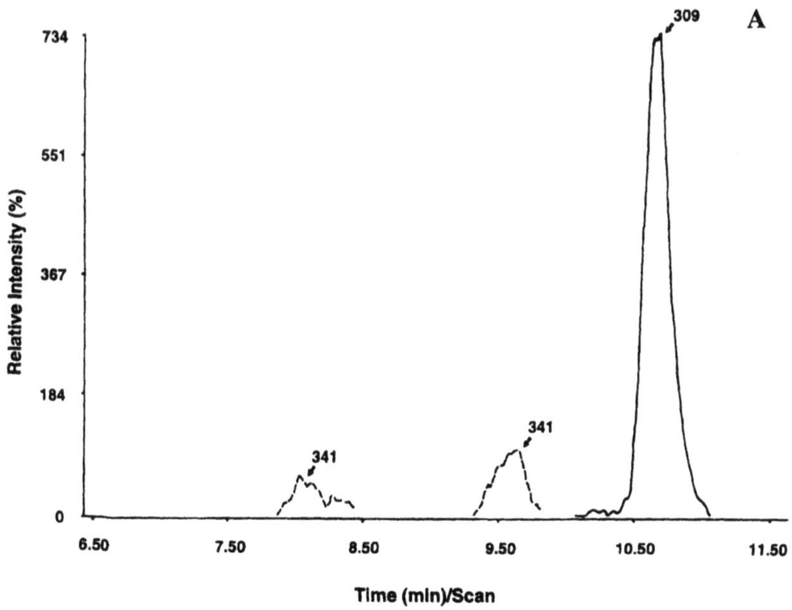

continued

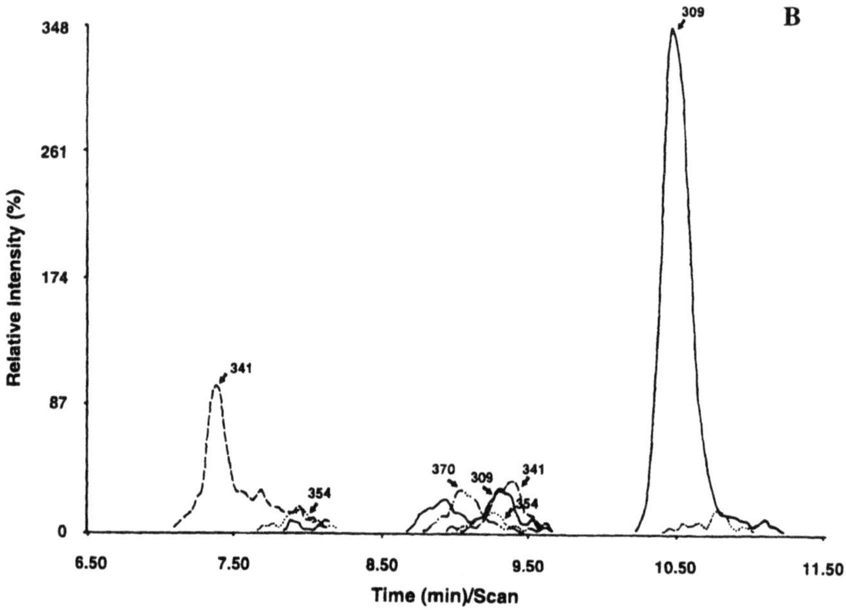

Figure 2. Liquid chromatography-mass spectroscopic analysis of xanthine oxidase-dependent linolenic acid oxidation products in the absence (**A**) and presence (**B**) of ·NO derived from S-nitroso-N-acetylpenicillamine. The ions detected represent [N-H]- ions. Liquid chromatography-mass spectroscopic analysis of reaction systems, performed immediately after addition of methanol to reaction systems to give a final concentration of 40% (v/v), revealed molecular compositional characteristics of novel derivatives produced by ·NO reaction with oxidizing linolenic acid. Linolenic acid oxidation induced by hypoxanthine, XO and EDTA-Fe^{+3} yielded two principal oxidation products, in addition to the [N-H]- ion (M/Z=277) of the native fatty acid. These species had a molecular mass of 310 (9- and/or 16-hydroperoxolinolenate) and 342 (9, 16-dihydroperoxolinolenate). Lower quantities of 9 and/or 16 hydroxylinolenate were also formed (molecular mass = 294). In lipid oxidation reactions simultaneously exposed to ·NO derived from SNAP, two additional ions were observed for novel products having molecular masses of 355 and 371, termed hydroxylnitrosoperoxolinolenate and hydroperoxonitrosolinolenate, respectively. Adapted from (Rubbo, 1994) with permission.

·NO and α-tocopherol not that dissimilar. From this, one might speculate that ·NO may, under some conditions, predominate over α-topcopherol as an oxidant-protective molecule in tissues. These reactions are especially important to consider, inasmuch as ·NO significantly concentrates in lipophilic cell compartments, having a lipid:water partition coefficient of 8:1. This will further enhance the ability of ·NO to serve as a critical regulator of oxidant-induced membrane or lipoprotein oxidation via termination of chain propagating radical reactions.

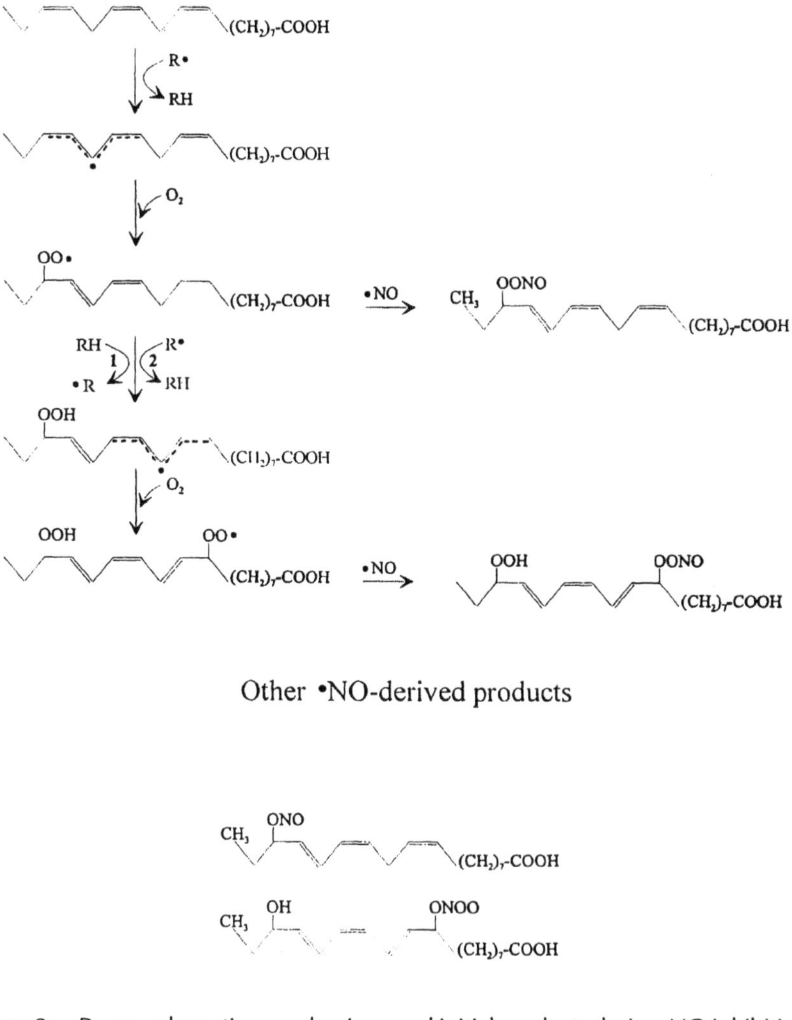

Figure 3. Proposed reaction mechanisms and initial products during ·NO inhibition of lipid oxidation. From (Rubbo, 1994) with permission.

C. Nitric Oxide Reactions with ·OH or *trans* $ONOO^-$

For kinetic reasons and based on recent experimental evidence, ·NO is expected to have only limited potential as a ·OH scavenger in both aqueous and lipid phases (Rubbo, 1994b). Nitric oxide reacts rapidly with ·OH ($k = 10 \times 10^{10}\ M^{-1} \cdot s^{-1}$) to form nitrous acid (Ross, 1992). The potent nonselective reactivity and short diffusion distance of ·OH would mean that there would have to be extremely high and non-biological concentrations of ·NO present to effectively compete with other target molecule reactions of ·OH. Experimentally, the ·OH scavengers mannitol and DMSO were observed to afford only minimal inhibition of $ONOO^-$-dependent hydroxylation of benzoate to salicylic acid (Table 3). This indicates that a) at the submicromolar concentrations of ·NO occurring in vivo and b) at ·NO concentrations ideally employed in reaction systems designed to reflect biological events, ·NO would not significantly inhibit the ·OH-like reactivity of $ONOO^-$ upon protonation to ONOOH. Since the concentration of benzoate was three orders of magnitude greater than ·NO in the aforementioned reaction systems, ·NO would not be expected to effectively compete with benzoate as a ·OH scavenger. The partial inhibitory effect of ·NO observed towards benzoate hydroxylation may also be due to reaction of ·NO with the *trans-* form of ONOOH, a conformation critical for expression of the ·OH-like reactivity of $ONOO^-$ following protonation (Crow, 1994).

The experimental conditions where only partial inhibition of benzoate hydroxylation by either nitrosothiols or pure ·NO gas dissolved in solution occurred were those which led to extensive inhibition of lipid oxidation. In lipid-containing reactions, fatty acids will also effectively compete with ·NO for reaction with the highly reactive ·OH. In aggregate, these observations indicate that the principal mechanism of ·NO inhibition of lipid oxidation is not via scavenging of $ONOO^-$ or ONOOH, rather it occurs by ·NO annihilation of lipid radical species, thus terminating radical chain propagation processes.

D. Nitric Oxide "Diversionary Reactions" with Thiols and Superoxide

Thiols are key biological antioxidants due to their redox properties, ubiquitous presence and high intra- and extracellular concentrations (i.e., the 1–10 mM glu-

Table 3. The Effect of Nitric Oxide and Hydroxyl Radical Scavengers on Peroxynitrite-Dependent Benzoate Hydroxylation.

Condition	*% Control relative fluorescence*
$ONOO^-$	145 ± 18
SNAP	26 ± 5
$ONOO^-$, SNAP	89 ± 11
$ONOO^-$, acetylpenicillamine	118 ± 19
·NO	4 ± 1
$ONOO^-$, ·NO	90 ± 9
$ONOO^-$, mannitol	53 ± 7
$ONOO^-$, mannitol, SNAP	25 ± 5
$ONOO^-$, DMSO	71 ± 9
$ONOO^-$, DMSO, SNAP	14 ± 4
$ONOO^-$, DTPA	152 ± 22
$ONOO^-$, DTPA, SNAP	93 ± 16

Notes: A continuous infusion of $ONOO^-$ (6.7 $\mu M.min^{-1}$) into a solution of 10 mM benzoate in 50 mM potassium phosphate, pH 6.8, 18° C was performed. The reaction mixtures contained (as indicated) 100 µM SNAP, 4.3 $\mu M.min^{-1}$ .NO, 200 µM DTPA, 50 mM mannitol or 50 mM dimethyl sulfoxide. Infusions were for 15 m and then analyzed for benzoate hydroxylation to salicylic acid via flourescence spectroscopy (λ_{ex} = 300 nm and (λ_{em} = 410 nm). Data represents mean ± SD, n=5. Adapted from (Rubbo, 1994) with permission.

tathione and the 0.4–0.5 mM albumin-SH concentrations found in cells and plasma, respectively). It is possible that ·NO indirectly protects thiols from irreversible oxidation by contributing to the formation of the less oxidant-sensitive S-nitrosothiol derivative. While this adduct is one which may lead to ADP-ribosylation, it is at the same time resistant to irreversible oxidation to the sulfenic and sulfonic acid derivatives. Thiols readily react with oxidants including ·OH, H_2O_2, O_2-, $\cdot NO_2$ and $ONOO^-$ (Radi, 1991a, 1991c; Ross, 1992; Meyer, 1994), with the oxidation of thiols by $ONOO^-$ representing a major pathway of biological $ONOO^-$ reactivity (Radi, 1993). While $\cdot NO_2$ and $ONOO^-$ react efficiently with thiols, ·NO does not readily react directly at neutral pH. Most of the thiol reactivity of ·NO is due to either a) the intermediate formation of $\cdot NO_2$, b) one electron oxidation of the thiol to the thiyl radical, which then can undergo S-nitrosation by ·NO, c) one electron oxidation of ·NO to the nitrosonium cation (NO+) which then can react with the thiolate to give RSNO or d) $ONOO^-$-mediated one electron oxidation of thiols to the thiyl radical, which then could react with adventitious ·NO. The precise chemistry of biological S-nitrosation has not been clarified yet, and is an area of intense scrutiny because of the therapeutic potential of ·NO-donating drugs. S-nitrosothiols are more stable than ·NO under aerobic conditions, with half-lives for spontaneous aerobic decomposition ranging from minutes (S-nitrosocysteine) to hours (S-nitrosoglutathione)

Nitric oxide reaction with $O_2^{\bullet -}$ may in some cases also serve to protect $O_2^{\bullet -}$ sensitive target molecules. This "diversionary reaction" of ·NO, which kinetically may outcompete superoxide dismutases, forces $O_2^{\bullet -}$ through $ONOO^-$ oxidation and decomposition pathways. At the same time, this will limit the accumulation of H_2O_2 and decrease the formation of secondary reactive species derived from H_2O_2 reaction with transition metals. In reaction systems and tissue culture models exposed to xanthine oxidase concentrations found in the human circulation following surgical procedures (0.5 to 2.5 mU XO/ml plasma) and sub-micromolar/min ·NO generation rates (derived from S-nitrosothiols), we have observed that ·NO induces a 40% reduction in the control (no added S-nitrosothiol) rate of H_2O_2 accumulation (unpublished observations). This event may promote the extracellular decomposition or alternative reactions of the less stable $ONOO^-$, while at the same time limiting the accumulation and reactions of H_2O_2, toxic to tissues in its own right.

E. Nitric Oxide Induction of cGMP-Dependent Protective Effects

Nitric oxide is a potent endogenous nitrovasodilator, playing a major role in modulating vascular tone. This paracrine factor exerts its vasoactive effect by stimulating soluble guanylate cyclase in the underlying vascular smooth muscle cells, thereby elevating intracellular levels of cGMP and inducing relaxation of the vascular smooth muscle (Moncada, 1991b). Nitric oxide also inhibits interaction of the vessel wall with circulating blood elements (ie platelets and neutrophils). The effect of ·NO in inhibiting platelet adhesion and aggregation to the vessel wall is associated with increases in the level of platelet cGMP. For neutrophils, the mechanism underlying ·NO-mediated reduction in vessel wall binding and the role of cGMP is less clear, and probably also involves modulation of vessel wall integrin function and gene expression (Kubes, 1994; Khan, 1995).

VI. CONCLUSIONS

In biological systems where co-generation of multiple reactive species occurs, ·NO will exacerbate oxidant injury via production of the potent oxidant $ONOO^-$ and can also exert tissue-protective roles. This can occur by a) redirection of $O_2^{\bullet -}$-mediated cytotoxic reactions to other oxidative pathways, b) inhibition of $ONOO^-$-dependent oxidation reactions, c) induction of a salutary cGMP-dependent cell metabolic and structural state and d) termination of free radical-dependent chain propagation reactions catalyzed by diverse initiating species. The relative rates of production, sites of production and steady state concentrations of reactive species, antioxidants and tissue mediators will critically influence the observed apparent toxic or protective effects of ·NO in biological systems. The cellular and anatomical sites of production of $O_2^{\bullet -}$ and ·NO, and the dominant operative mechanisms of oxidant damage in tissues at the time of $O_2^{\bullet -}$ and ·NO production profoundly influence expression of the differential oxidant injury-enhancing and protective effects of ·NO. Development of a better understanding of the physiological roles of ·NO, coupled with detailed insight into ·NO regulation of oxygen radical-dependent reactions and the chemistry of ·NO and $ONOO^-$, should yield a more rational basis for the present and future therapeutic use of inhaled ·NO gas mixtures, ·NO donors and inhibitors of ·NO synthases.

REFERENCES

Beckman, J.S., Beckman, T.W., Chen, J., Marshall, P.A., & Freeman, B.A. (1990). Apparent hydroxyl radical production by peroxynitrite: Implications for endothelial injury from nitric oxide and superoxide. Proc. Natl. Acad. Sci. USA 87,1620-1624.

Beckman, J.S. (1991). The double edged role of nitric oxide in brain function and superoxide-mediated pathology. J. Devel. Physiol. 15, 53-59.

Beckman, J.S., Ye, Y.Z., Anderson, P.G., Chen, J., Accavitti, M.A., Tarpey, M.M., & White, R. (1994). Extensive nitration of protein tyrosines in human atherosclerosis detected by immunohistochemistry. Biol. Chem. Hoppe-Seyler 375, 81-88.

Bredt, D.S., Hwang, P.M., Glass, C.E., Lowenstein, C., Reed, R., & Snyder, S.H. (1991). Cloned and expressed nitric oxide synthase structurally resembles cytochrome P-450 reductase. Nature 35, 714-718.

Carreras, M.C., Pargament, G.A., Catz, S.D., Poderoso, J.J., & Boveris, A. (1994). Kinetics of nitric oxide and hydrogen peroxide production and formation of peroxynitrite during the respiratory burst of human neutrophils. FEBS Lett. 341, 65-68.

Caselli, A., Camici, G., Manao, G., Moneti, G., Pazzagli, L., Cappugi, G., & Ramponi, G. (1994). Nitric oxide causes inactivation of the low molecular weight phosphotyrosine protein phosphatase. J. Biol. Chem. 269, 24878-24882.

Castro, L., Rodriguez, M., & Radi, R. (1994). Aconitase is readily inactivated by peroxynitrite, but not by its precursor, nitric oxide. J. Biol. Chem. 269, 29409-29415.

Choi, D.W. (1993). Nitric oxide: Foe or friend to the injured brain?. Proc. Natl. Acad. Sci. USA 90, 9741-9743.

Clancy, R.M., Leszczynska, J., & Abramson, S.B. (1992). Nitric oxide, an endothelial cell relaxation factor, inhibits neutrophil superoxide anion production via a direct action on the NADPH oxidase. J. Clin. Invest. 90, 1116-1121.

Cleeter, M.W.J., Cooper, J.M., Darley-Usmar, V.M., Moncada, S., & Schapira, A.H.V. (1994). Reversible inhibition of cytochrome c oxidase, the terminal enzyme of the mitochondrial respiratory chain, by nitric oxide. Implications for neurodegenerative diseases. FEBS Lett. 345, 50-54.

Crow, J.P., Spruell, C., Chen, J., Gunn, C., Ischiropoulos, H., Tsai, M., Smith, C., Radi, R., Koppenol, W.H., & Beckman, J.S. (1994). On the pH-dependent yield of hydroxyl radical products from peroxynitrite. Free Rad. Biol. Med. 16, 331-338.

Denicola, A., Rubbo, H., Rodriguez, D., & Radi, R. (1993). Peroxynitrite-mediated cytotoxicity to *Trypanosoma cruzi*. Arch. Biochem. Biophys. 304, 279-286.

Drapier, J., Hirling, H., Wietzerbin, J., Kaldy, P., and Kuhn, L.C. (1993). Biosynthesis of nitric oxide activates iron regulatory factor in macrophages. EMBO J. 12, 3643-3649.

Drexler, H., Zeiher, A., Meinzer, K., & Just, H. (1991). Correction of endothelial dysfunction in coronary microcirculation of hypercholesterolaemic patients by L-arginine. Lancet 338, 1546-1550.

Drexler, H., Fischell, T., Pinto, F., Chenzbraun, A., Botas, J., Cooke, J., & Alderman, E.L. (1994). Effect of L-arginine on coronary endothelial function in cardiac transplant recipients. Circulation 89, 1615-1623.

Edwards, J.O., & Plumb, R.C. (1993). The chemistry of peroxonitrites. Prog. Inorg. Chem. 41, 599-635.

Fukahori, M., Ichimori, K., Ishida, H., Nakagawa, H., & Okino, H. (1994). Nitric oxide reversibly suppresses xanthine oxidase activity. Free Rad. Res. 21, 203-212.

Furchgott, R.F., & Zawadski, J.B. (1980). The obligatory role of endothelial cells in the relaxation of arterial smooth muscle by acetylcholine. Nature 288, 373-376.

Graham, A., Hogg, N., Kalyanaraman, B., O'Leary, V., Darley-Usmar, V., & Moncada, S. (1993). Peroxynitrite modification of low-density lipoprotein leads to recognition by the macrophage scavenger receptor. FEBS Lett. 330, 181-185.

Guo, J.P., Siegfried, M.R., & Lefer, A.M. (1994). Endothelial preserving actions of a nitric oxide donor in carotid arterial intimal injury. Methods Find. Exp. Clin. Pharmacol. 16, 347-354.

Haddad, I.Y., Ischiropoulos, H., Holm, B.A., Beckman, J.S., Baker, J.R. & Matalon, S. (1993). Mechanism of peroxynitrite-induced injury to pulmonary surfactants. Am. J. Physiol 265, L555-L564.

Haddad, I.Y., Pataki, G., Hu, P., Galliani, C., Beckman, J.S., & Matalon, S. (1994). Quantitation of nitrotyrosine levels in lung sections of patients and animals with acute lung injury. J. Clin. Invest. 94, 2407-2413.

Hibbs, J.B., Taintor, R.R., Vavrin, Z., & Rachlin, E.M. (1988). Nitric oxide: A cytotoxic activated macrophage effector molecule. Biochem. Biophys. Res. Comm. 157, 87-94.

Hogg, N., Darley-Usmar, V., Graham, A., & Moncada, S. (1993a). Peroxynitrite and atherosclerosis. Biochem. Soc. Trans 21, 358-361.

Hogg, N., Kalyanaraman, B., Joseph, J., Struck, A., & Parthasarathy, S. (1993b). Inhibition of low-density lipoprotein oxidation by nitric oxide. Potential role in atherogenesis. FEBS Lett. 334, 170-174.

Hu, P., Ischiropoulus, H., Beckman, J.S., & Matalon, S. (1994). Peroxynitrite inhibition of oxygen consumption and sodium transport in alveolar type II cells. Am. J. Physiol. 266, L628-L634.

Huie, R.E., & Padmaja, S. (1993). Reaction of NO with $O_2^{\bullet -}$. Free Rad. Res. Comm. 18,195-199.

Ischiropoulos, H., Zhu, L., Chen, J., Tsai, M., Martin, J., Smith, C., & Beckman, J.S. (1992a). Peroxynitrite-mediated nitration of tyrosine catalyzed by superoxide dismutase. Arch. Biochem. Biophys. 298, 431-437.

Ischiropoulos, H., Zhu, L., & Beckman, J.S. (1992b). Peroxynitrite formation from activated rat alveolar macrophages. Arch. Biochem. Biophys. 298, 446-451.

Jessup, W., & Dean, R.T. (1993). Autoinhibition of murine macrophage-mediated oxidation of low-density lipoprotein by nitric oxide synthesis. Atherosclerosis 101, 145-155.

Kanner, J., Harel, S., & Granit, R. (1991). Nitric oxide as an antioxidant. Arch. Biochem. Biophys. 289,130-136.

Kanner, J., Harel, S., & Granit, R. (1992). Nitric oxide, an inhibitor of lipid oxidation by lipoxygenase, cyclooxygenase and hemoglobin. Lipid 27, 46-49.

Kavanagh, B.P., Mouchawar, A., Goldsmith, J., & Pearl, R.G. (1994). Effects of inhaled nitric oxide and inhibition of endogenous nitric oxide synthesis in oxidant-induced acute lung injury. J. Appl. Physiol. 76, 1324-1329.

Kelm, M., & Schrader, J. (1990). Control of coronary vascular tone by nitric oxide. Circ. Res. 66, 1561-1575.

Khan, B.V., Parthasarathy, S., Alexander, R.W., & Medford, R.M. (1995). Modified low density lipoprotein and its constituents augment cytokine-activated VCAM-1 gene expression in human vascular endothelial cells. J. Clin. Invest. (In Press).

Kharitonov, V.G., Sundquist, A.R., & Sharma, V.S. (1994). Kinetics of nitric oxide autooxidation in aqueous solution. J. Biol. Chem. 269, 5881-5883.

King, P.A., Jamison, E., Strahs, D., Anderson, V.E., & Brenowitz, M. (1994). Footprinting proteins on DNA with peroxonitrous acid. Nucleic Acids Res. 21, 2473-2478.

Knowles, R.G., & Moncada, S. (1993). Nitric oxide as a signal in blood vessels. Trends Biochem. Sci. 17, 39-402

Kooy, N.W., & Royall, J.A. (1994). Agonist-induced peroxynitrite production from endothelial cells. Arch. Biochem. Biophys. 310, 352-359.

Koppenol, W.H., Moreno, J.J., Pryor, W.A., Ischiropoulos, H., & Beckman, J.S. (1992). Peroxynitrite, a cloaked oxidant formed by nitric oxide and superoxide. Chem. Res. Toxicol. 5, 834-842.

Kubes, P. (1993). Ischemia-reperfusion in feline small intestine: A role for nitric oxide. Am. J. Physiol. 264, G143-G149.

Kubes, P. (1994). The microcirculation and inflammation: modulation of leukocyte-endothelial cell adhesion. J. Leuk. Biol. 55, 662-675.

Kubes, P., Suzuki, M., & Granger, D.N. (1991). Nitric oxide: An endogenous modulator of leukocyte adhesion. Proc. Natl. Acad. Sci. USA 88, 4651-4655.

Kurose, I., Wolf, R., Grisham, M.B., & Granger, D.N. (1994). Modulation of ischemia/reperfusion-induced microvascular dysfunction by nitric oxide. Circ. Res. 74, 376-382.

Lefer, D.J., Nakanishi, K., & Vinten-Johansen, J. (1993). Endothelial and myocardial cell protection by a cysteine-containing nitric oxide donor after myocardial ischemia and reperfusion. J. Cardiovasc. Pharmacol. 22(Suppl. 7), S34-S43.

Lepoivre, M., Flaman, J., Bobe, P., Lemaire, G., & Henry, Y. (1994). Quenching of the tyrosyl free radical of ribonucleotide reductase by nitric oxide: Relationship to cytostasis induced in tumor cells by cytotoxic macrophages. J. Biol. Chem. 269, 21891-21897.

Mahoney, L.R. (1970). Evidence for the formation of hydroxyl radical in the isomerization of pernitrous acid to nitric acid in aqueous solution. J. Am. Chem. Soc. 92, 5262-5263.

Malo-Ranta, U., Yla-Herttuala, S., Metsa-Ketela, T., Jaakkola, O., Moilanen, E., Vuorinen, P., & Nikkari, T. (1994). Nitric oxide donor GEA 3162 inhibits endothelial cell-mediated oxidation of low density lipoprotein. FEBS Lett. 337, 179-183.

Marsden, P.A, Schappert, K.T., Chen,, H.S., Flowers, M., Sundell, C.L., Wilcox, J.N., Lamas, N., & Michel, T. (1992). Molecular cloning and characterization of human endothelial nitric oxide synthase. FEBS Lett 307, 287-293.

Matheis, G., Sherman, M.P., Buckberg, G.D., Haybron, D.M., Young, H.H., & Ignarro, L.J. (1992). Role of L-arginine-nitric oxide pathway in myocardial reoxygenation injury. Am. J. Physiol. 262,H616-H620.

Meyer, D.J., Kramer, H., Ozer, N., Coles, B., & Ketterer, B. (1994). Kinetics and equilibria of S-nitrosothiol-thiol exchange between glutathione, cysteine, penicillamines and serum albumin. FEBS Lett. 345, 177-180.

Mohr, S., Stamler, J.S., & Brune, B. (1994). Mechanism of covalent modification of glyceraldehyde-3-phosphate dehydrogenase at its active site thiol by nitric oxide, peroxynitrite and related nitrosating agents. FEBS Lett. 348, 223-227.

Moncada, S., & Higgs, E.A. (1991a). Endogenous nitric oxide: physiology, pathology and clinical relevance. Eur. J. Clin. Inv. 21, 361-374.

Moncada, S., Palmer, R.M.J., & Higgs, E.A. (1991b). Nitric oxide: physiology, pathology, and pharmacology. Pharmacol. Rev. 43, 109-142.

Moro, M.A., Darley-Usmar, V., Goodwin, D., Read, N., Zamora-Pino, R., Feelisch, M., Radomski, M., & Moncada, S. (1994). Paradoxical fate and biological action of peroxynitrite on human platelets. Proc. Natl. Acad. Sci. USA 91, 6702-6706.

Mulligan, M.S., Hevel, J.M., Marletta, M.A., & Ward, P.A. (1991). Tissue injury caused by deposition of immune complexes is L-arginine dependent. Proc. Natl. Acad. Sci. USA 88, 6338-6342.

Mulligan, M.S., Warren, J.S., Smith, C.W., Anderson, D.C., Yeh, C.G., Rudolph, A.R., & Ward, P.A. (1992). Lung injury after deposition of Ig A immune complexes. Requirements for CD18 and L-arginine. J. Immunol. 148, 3086-3092.

Ohara, A., Gatti, R.M., & Radi, R. (1994). Spin-trapping studies of peroxynitrite decomposition and of 3-morpholinosydonimine N-ethylcarbamide autooxidation: Direct evidence for metal-independent formation of free radical intermediates. Arch. Biochem. Biophys. 310, 118-125.

Padmaja, S., & Huie, R.E. (1993). The reaction of nitric oxide with organic peroxyl radicals. Biochem. Biophys. Res. Comm. 195, 539-544.

Palmer, R.M.J., Ferrige, A.G., & Moncada, S. (1987). Nitric oxide release accounts for the biological activity of endothelium-derived relaxing factor. Nature 327, 524-526.

Payne, D., & Kubes, P. (1993). Nitric oxide donors reduce the rise in reperfusion-induced intestinal mucosal permeability. Am. J. Physiol. 265, G189-G195.

Pinsky, D.J., Naka, Y., Chowdhury, N., Liao, H., Oz, M.C., Michler, R.E., Kubaszewski, E., Malinski, T., & Stern, D.M. (1994). The nitric oxide/cyclic GMP pathway in organ transplantation: Critical role in successful lung preservation. Proc. Natl. Acad. Sci. USA 91, 12086-12090.

Pryor, W.A., Jin, X., & Squadrito, G.L. (1994). One-and two- electron oxidations of methionine by peroxynitrite. Proc. Natl. Acad. Sci. USA 91, 11173-11177.

Radi, R., Beckman, J.S, Bush, K., & Freeman, B.A. (1991a). Peroxynitrite oxidation of sulfhydryls: the cytotoxic potential of endothelial-derived superoxide and nitric oxide. J. Biol. Chem. 266, 4244-4250.

Radi, R., Beckman, J.S, Bush, K., & Freeman, B.A. (1991b). Peroxynitrite-induced membrane lipid peroxidation: the cytotoxic potential of superoxide and nitric oxide. Arch. Biochem. Biophys. 288,481-487.

Radi, R., Bush, K., Cosgrove, T., & Freeman, B.A. (1991c). Reaction of xanthine oxidase-derived oxidants with lipid and protein of human plasma. Arch. Biochem. Biophys. 286, 117-125.

Radi, R. (1993). Oxidative chemistry of peroxynitrite: direct attack vs. the hydroxyl radical-like pathway. Free Rad. Biol. Med. 15, 494.

Radi, R., Rodriguez, M., Castro, L., & Telleri, R. (1994). Inhibition of mitochondrial electron transport by peroxynitrite. Arch. Biochem. Biophys. 308,89-95.

Reif, D.W., & Simmons, R.D. (1990). Nitric oxide mediates iron release from ferritin. Arch. Biochem. Biophys. 283, 537-541.

Ross, A.B., Mallard, W.G. & Bielski, B.H.J. (1992). NDRL-NIST solution kinetics database Version 1. National Institute of Standards and Technology, Gaithersburg, MD.

Rossaint, R., Gerlach, H., & Falke, K.J. (1994). Inhalation of nitric oxide- a new approach in severe ARDS. Eur. J. Anaesthesiol. 11, 43-51.

Rubbo, H., Denicola, A., & Radi, R. (1994a). Peroxynitrite inactives thiol-containing enzymes of *T. cruzi* oxidative metabolism and inhibits cell respiration. Arch. Biochem. Biophys. 308, 96-102.

Rubbo, H., Radi, R., Trujillo, M., Telleri, R., Kalyanaraman, B., Barnes, S., Kirk, M., & Freeman, B.A. (1994b). Nitric oxide regulation of superoxide and peroxynitrite-dependent lipid peroxidation: Formation of novel nitrogen-containing oxidized lipid derivatives. J. Biol. Chem. 269, 26066-26075.

Rubbo, H., Parthasarathy, S., Kalyanaraman, B., Barnes, S., Kirk, M., & Freeman, B.A. (1995). Nitric oxide inhibition of lipoxygenase-dependent liposome and low density lipoprotein oxidation: Formation of novel nitrogen-containing oxidized lipid derivatives. J. Biol. Chem, In Press.

Scharfstein, J.S., Keaney, J.F., Slivka, A., Welch, G.N., Vita, J.A., & Stamler, J.S. (1994). *In vivo* transfer of nitric oxide between a plasma protein-bound reservoir and low molecular weight thiols. J. Clin. Invest. 94, 1432-1439.

Schweizer, M., & Richter, C. (1994). Nitric oxide potently and reversibly deenergizes mitochondria at low oxygen tension. Biochem. Biophys. Res. Comm. 204, 169-175.

Siegfried, M.R., Carey, C., Ma, X., & Lefer, A.M. (1992). Beneficial effects of SPM-5185, a cysteine-containing nitric oxide donor in myocardial ischemia-reperfusion. Am. J. Physiol. 263, H771-H777.

Snyder, S.H. Nitric oxide: first in a new class of neurotransmitters? (1993). Science 257, 494-496.

Stamler, J.S., Simon, D.I., Osborne, J.A., Mullins, M.E., Jaraki, O., Michel, T., Singel, D., & Loscalzo, J. (1992a). S-nitrosylation of proteins with nitric oxide: Synthesis and characterization of biologically active compounds. Proc. Natl. Acad. Sci. USA 89, 444-448.

Stamler, J.S., Singel, D.J., & Loscalzo, J. (1992b). Biochemistry of nitric oxide and its redox-activated forms. Science 258, 1898-1903.

Traylor, T.G., & Sharma, V.S. (1992). Why NO?. Biochemistry 31, 2847-2849.

van der Vliet, A., O'Neill, C., Halliwell, B, Cross, C.E., & Kaur, H. (1994a). Aromatic hydroxylation and nitration of phenylalanine and tyrosine by peroxynitrite. FEBS Lett. 339, 89-92.

van der Vliet, A., Smith, D., O'Neill, C., Kaur, H., Darley-Usmar, V., Cross, C.E., & Halliwell, B. (1994b). Interactions of peroxynitrite with human plasma and its constituents: oxidative damage and antioxidant depletion. Biochem. J. 303, 295-301.

Vanderkooi, J.M., Wright, W.W., & Erecinska, M. (1994). Nitric oxide diffusion coefficients in solutions, proteins and membranes determined by phosphorescence. Biochim. Biophys. Acta 1207, 249-254.

Villa, L.M., Salas, E., Darley-Usmar, V., Radomski, M., and Moncada, S. (1994). Peroxynitrite induces both vasodilatation and impaired vascular relaxation in the isolated perfused rat heart. Proc. Natl. Acad. Sci. USA 91, 12383-12387.

White, R., Brock, T., Chang, L., Crapo, J., Briscoe, P., Ku, D., Bradley, W., Gianturco, S., Gore, J., Freeman, B.A., & Tarpey, M.M. (1994). Superoxide and peroxynitrite in atherosclerosis. Proc. Natl. Acad. Sci. USA 91, 1044-1048.

Wink, D.A., Osawa, Y., Darbyshire, J., Jones, C.R., Eshenaur, S., & Nims, R.W. (1993a). Inhibition of cytochromes P450 by nitric oxide and a nitric oxide-releasing agent. Arch. Biochem. Biophys. 300, 115-123.

Wink, D.A., Hanbauer, I., Krishna, M.C., DeGraff, W., Gamson, J., & Mitchell, J.B. (1993b). Nitric oxide protects against cellular damage and cytotoxicity from reactive species. Proc. Natl. Acad. Sci. USA 90, 9813-9817.

Wu, M., Pritchard, K.A., Kaminski, P.M., Fayngersh, R.P., Hintze, T.H., & Wolin, M.S. (1994). Involvement of nitric oxide and nitrosothiols in relaxation of pulmonary arteries to peroxynitrite. Am. J. Physiol. 266, H2108-H2113.

Yao, S., Ober, J., Krishnaswami, A., Ferguson, J., Anderson, H., Golino, P., Buja, L., & Willerson, J.T. (1992). Endogenous nitric oxide protects against platelet aggregation and cyclic flow variations in stenosed and endothelium-injured arteries. Circulation, 86, 1302-1309.

Zhu, L., Gunn, C., & Beckman, J.S. (1992). Bactericidal activity of peroxynitrite. Arch. Biochem. Biophys. 298, 452-457.

Zweier, J.L., Kuppusamy, P., Williams, R., Rayburn, B., Smith, D., Weisfeldt, M.L., & Flaherty, J.T. (1989). Measurement and characterization of postischemic free radical generation in the isolated perfused heart. J. Biol. Chem. 264, 18890-18895.

IRON, OXYGEN RADICALS, AND DISEASE

Sally K. Nelson and Joe M. McCord

I. INTRODUCTION

An adequate supply of iron is critical to good health, and this has been recognized for many decades. The essential biochemical roles of iron have likewise been well-

Advances in Molecular and Cellular Biology
Volume 25, pages 157-183.

ISBN: 0-7623-0379-4

studied and understood for decades. Unfortunately, this knowledge has reinforced a common maxim of conventional wisdom: "If a little is good, more is better." The maxim is probably wrong in most cases; it is certainly wrong in the case of iron. In the past 20 years or so, many studies have implicated high storage levels of iron as a risk factor for heart disease (Sullivan, 1981, 1992, 1989), cancer (Selby and Friedman, 1988; Stevens et al., 1988; Milman et al., 1991), and many other diseases. In the same time frame, much new biochemical information has come to light. It suggests that once the biochemical requirements for iron have been met, any additional iron absorbed by the body serves no useful function, but rather poses a substantial liability in times of oxidative stress. The free radical production associated with a broad spectrum of disease states can mobilize excess iron from its storage protein, ferritin. Once liberated, this redox-active metal is a powerful initiator of lipid peroxidation, that greatly amplifies damage to cells and tissues. This chapter will therefore focus on the role of iron in oxidative damage and the impact that iron status has on disease processes. We will also consider whether the difference in iron stores seen between men and women has any relationship to the incidence of certain other diseases.

II. IRON, LIPID PEROXIDATION, AND OXIDATIVE STRESS

The role of iron as a catalyst in lipid peroxidation and cellular damage is well accepted. The possibility that the ready availability of iron may exacerbate or amplify many mechanisms of disease, especially those involving oxidative injury, is now an area of intense interest. Higher organisms are particularly careful about how iron is handled. There is never any "free" or loosely chelated iron in the healthy state. It is transported in the ferric state, bound to transferrin, in a complex that is especially difficult to reduce. Practically all the cellular iron is stored in the ferric state by ferritin, a protein found in virtually all tissues and in plasma. Ferritin and hemosiderin, which is thought to be an insoluble and possibly degraded form of ferritin, are therefore the principal proteins involved in the storage of iron. Iron mediated exacerbation of oxidative stress is well documented (Van der Kraaij et al., 1988; Halliwell and Gutteridge, 1988; Halliwell and Gutteridge, 1986; Halliwell and Gutteridge, 1985; Braughler et al., 1986). The superoxide radical has been shown to bring about the reductive release of iron from ferritin (Biemond et al., 1984, 1986; Bolann and Ulvik, 1987; Thomas et al., 1985):

$$O_2^{\bullet -} + \text{Ferritin}(Fe^{+3}) \rightarrow O_2 + \text{Ferritin} + Fe^{+2}$$

It is this iron, liberated by the pathological production of superoxide, that is now free to initiate lipid peroxidation or to catalyze Haber-Weiss chemistry resulting in the formation of HO+ (McCord and Day, 1978):

$$O_2^{\bullet -} + Fe^{+3} \rightarrow O_2 + Fe^{+2}$$
$$Fe^{+2} + H_2O_2 \rightarrow Fe^{+3} + OH^- + HO+$$
$$\overline{O_2^{\bullet -} + H_2O_2 \rightarrow O_2 + OH^- + HO+}$$

The HO+ radical is capable of abstracting a hydrogen atom from polyunsaturated fatty acids (LH) to initiate lipid peroxidation:

$$HO+ + LH \rightarrow H_2O + L+$$
$$L+ + O_2 \rightarrow LOO\cdot$$
$$LOO\cdot + LH \rightarrow LOOH + L+$$

Once lipid hydroperoxides (ROOH) accumulate, free iron may directly initiate additional lipid peroxidation:

$$Fe^{2+} + LOOH \rightarrow Fe^{3+} + LO\cdot + OH^-$$
$$LO\cdot + LH \rightarrow LOH + L+, \text{etc.}$$

The resulting accumulation of lipid hydroperoxides destroys membrane structure and function. Because iron plays an important role in the amplification of oxidative injury, readily available or excessively high physiological stores of iron represent a substantial liability to an organism in oxidative stress, when oxygen-derived radical production is also high. This effect is seen in hemochromatosis where the excess tissue iron can be associated with congestive heart failure (Easley et al., 1972).

A. Iron Release from Stores

As iron plays a crucial role in oxidative stress, it is important to understand the sensitivity of the storage forms of iron to the presence of oxygen free radicals. Ferritin is the main protein involved in the storage of iron. Superoxide ion has been shown to bring about release of iron from ferritin (Biemond et al., 1984, 1986; Bolann and Ulvik, 1987, 1990; Thomas et al., 1985). It has been proposed (Biemond et al., 1988) that $O_2^{\bullet -}$ enters the ferritin-core through the hydrophilic channels, followed by reduction of ferric iron to the ferrous state. This enables the release of iron from the ferritin-core such that it may then catalyze the formation of HO+ or initiate lipid peroxidation. A recent study shows that this superoxide-dependent iron release from ferritin is species specific (Harris et al., 1994). The initial rates of iron release were much higher for ferritins from human, trout and rat liver than those from lamprey liver or horse spleen. Moreover, approximately 90% of the iron from the former group was ultimately releasable by superoxide, whereas a very small percentage of total iron could be released by superoxide from horse spleen ferritin. This information clears much of the confusion in the literature concerning the ability of superoxide to release iron from ferritin. Because of its ready commercial availability, many published studies have used horse spleen ferritin (Bolann and Ulvik, 1990).

A number of pathogenic or toxic agents other than superoxide can also effect the liberation of iron from ferritin. Exposure of ferritin to ionizing radiation (Reif et al., 1988), to redox-cycling xenobiotics such as paraquat (Thomas and Aust, 1986), adriamycin, alloxan (Reif, 1992), bleomycin (Norskov-Lauritsen et al., 1987), 6-hydroxydopamine (Monteiro and Winterbourn, 1989), or to other agents capable of forming reducing radicals (Monteiro et al., 1989) results in iron release. Aqueous extracts of tar from cigarette smoke can similarly cause the release (Moreno et al., 1992; Lapenna et al., 1995).

Whatever the agent that causes ferritin to release its iron, the mechanism must be a reductive one in order to liberate ferrous iron. No evidence suggests that physiologically relevant concentrations of H_2O_2 can release iron from ferritin. Balla et al. (1992) have suggested that holoferritin is cytoprotective to the endothelium during oxidative stress, but their model (addition of hydrogen peroxide to cultured cells) may not be an appropriate reflection of physiological oxidative stress. The superoxide radical, paradoxically, is a better reductant than it is an oxidant. Much of the *oxidative* damage brought about by superoxide production results from its *reductive* liberation and recycling of iron.

Reif and Simmons (1990) found that sodium nitroprusside, which decomposes to produce nitric oxide (NO+) causes iron release from ferritin. The mechanism for this reaction is unclear, but may involve the donation of an electron by NO+ to produce NO^+ and an atom of ferrous iron. Kanner et al. (1991), on the other hand, suggest that NO+ acts as an antioxidant with specific reference to iron-catalyzed reactions. Clearly, much more work needs to be done in this area. Another species to be considered is nitroxyl (NO^-), which has been suggested to be produced from the reversible interaction of nitric oxide with reduced superoxide dismutase (Murphy and Sies, 1991) as well as by nitric oxide synthase directly (Hobbs et al., 1994).

Although ferritin is considered the major source of iron that is liberated during oxidative stress, there may be additional sources. Hemosiderin is a virtually insoluble protein-iron complex, the origin of which is not clear. There is some evidence that hemosiderin is a polymerized (Hoy and Jacobs, 1981) or otherwise degraded (O'Connell et al., 1986) form of ferritin, and even a speculation that it may be a protective mechanism of sorts, as its iron seems to be more resistant to mobilization than that of ferritin (O'Connell et al., 1986). It seems more likely that it reflects accumulated oxidative damage within the cell, as is thought to be the case for lipofuscin (Sohal and Brunk, 1990).

Brieland and Fantone (1991) have shown phorbol myristate acetate (PMA)-stimulated neutrophils enhanced the release of ferrous iron from saturated transferrin at pH 7.4, in association with the generation of superoxide. Bannister et al. (1982b) have also demonstrated that hydroxyl radicals are generated in the presence of transferrin following superoxide production by neutrophil NADPH oxidase. Miller and Britigan (1995) found that a *Pseudomonas*-derived elastase acts on transferrin to render the iron redox active, capable of catalyzing hydroxyl radical formation. Whether neutrophil proteases can similarly degrade transferrin is an interesting question for future research. There have been contrasting reports on the role of lactoferrin in HO+ production (Britigan et al., 1989; Bannister et al., 1982a). Halliwell and Gutteridge (1986) do not believe lactoferrin or transferrin, at physiological levels of iron loading, could be promoters of the hydroxyl radical. The Fe^{3+} embedded in the two specific sites in these proteins cannot easily react to form the HO+ that could escape into free solution. However, if the proteins were partially degraded by neutrophil proteases, the reaction might be possible.

Oxidative degradation of hemoglobin with the release of iron has been shown in the presence of acrolein (Ferrali et al., 1989, 1990), phenylhydrazine, divicine, isouramil (Ferrali et al., 1992) and both H_2O_2 and organic hydroperoxides (Gutteridge, 1986). There is some evidence of a cellular pool of transient ferric iron that is chelatable by desferrioxamine (Rothman et al., 1992), which appears to be kinetically and immunochemically distinct from the larger cellular pool of ferritin iron. Desferrioxamine depletes the cellular pool of chelatable iron and protects cells from oxidative injury.

B. Iron in Ischemia and Reperfusion Injury

When the blood flow to a tissue is temporarily interrupted, metabolic changes occur that result in dramatic overproduction of superoxide radical when the tissue is once again perfused and reoxygenated (McCord, 1985). The burst of oxygen-derived free radicals in myocardial reperfusion injury has been demonstrated by electron paramagnetic resonance using spin-trapping techniques (Grill et al., 1992; Arroyo et al., 1987; Garlick et al., 1987; Bolli et al., 1988). As described above, the production of superoxide radical results in the liberation and reduction of iron from tissue ferritin (Biemond et al., 1984, 1986; Powers et al., 1992), as well as the secondary formation of H_2O_2 and hydroxyl radical via Haber-Weiss chemistry (McCord and Day, 1978; Rubanyi and Vanhoutte, 1986). Tissue iron is delocalized during reperfusion (Krause et al., 1985; Voogd et al., 1992), and is released from an ischemic organ to the vascular space (Voogd et al., 1992, 1993; Moss et al., 1988; Paller and Hedlund, 1988). Iron-overload exacerbates reperfusion injury (Van der Kraaij et al., 1988; Healing et al., 1990; Fuller et al., 1987; White et al., 1985). Desferrioxamine, a powerful chelator of ferric iron that prevents redox-cycling, significantly improves functional and metabolic recovery of isolated perfused rabbit hearts subjected to 30 minutes of global ischemia (Williams et al., 1991). Similarly, infarct size is significantly decreased in animals pretreated with the drug (Reddy et al., 1989; Lesnefsky et al., 1990). Rats treated with desferrioxamine doubled the probability of survival (from 31% to 62%) when subjected to total circulatory arrest, followed by resuscitation (Babbs, 1985). Desferrioxamine decreased release of creatine kinase in the isolated rabbit (Myers et al., 1985) or rat (Badylak et al.,

1987) heart subjected to hypoxia and reoxygenation and also prevented reperfusion-induced arrhythmias (Bernier et al., 1986). It substantially improved recovery from myocardial "stunning" in desferrioxamine-treated open-chest dogs (Bolli et al., 1987; Farber et al., 1988). It also improved post-ischemic recovery of high-energy phosphate content as measured by ^{31}P nuclear magnetic resonance spectroscopy (Bernard et al., 1988). A marked protection from cardiac injury in the isolated perfused rat hearts has been shown by neocuproine, a chelator for both iron and copper (Appelbaum et al., 1990).

Desferrioxamine may improve recovery of transplanted organs. Desferrioxamine added to the cardioplegia solution substantially improved postreperfusion ventricular pressure development, maximal rate of rise of ventricular pressure, left ventricular compliance, and coronary flow (Menasche et al., 1990). Cold ischemic rabbit kidneys were less susceptible to lipid peroxidation when treated with desferrioxamine (Green et al., 1986; Fuller et al., 1987). Liver subjected to warm ischemia and reperfusion *in vivo* is protected substantially from lipid peroxidation and by histopathological evidence if desferrioxamine is administered at any point prior to reperfusion (Omar et al., 1989).

Nutritional iron status may be a predisposing factor in ischemic heart disease. A study by van der Kraaij et al. (1988) has addressed the question by comparing hearts from normal and iron-supplemented rats in Langendorff preparation, subjected to 45 minutes of anoxic perfusion followed by reoxygenation. Under normoxia, the groups behaved identically. Following a period of anoxia, however, the iron-loaded hearts displayed ventricular fibrillation and significantly lower recovery of contractility, both of which were returned to control values by perfusion with desferrioxamine. Because iron and hydroxyl radical are both initiators of lipid peroxidation (Halliwell and Gutteridge, 1985; Aust et al., 1985), one might expect lipid peroxidation to be a prominent component of reperfusion injury. Many reports confirm this to be the case (Meerson et al., 1982; Tavazzi et al., 1992; Yamashita et al., 1992).

III. A WORD ABOUT NUTRITIONAL STATUS AND IRON STORES

Defining the "normal" range of any health-related parameter is risky business, and it may be especially so in the case of iron. "Normal" often simply reflects the average for the population at hand, and does not necessarily mean "optimal" or "desirable" or "healthy". For many decades the health consequences of iron deficiency have been apparent. True iron deficiency is easily and cheaply remedied by iron supplementation. Only in the last decade or so have the consequences of iron excess (i.e., substantial ferritin stores) been recognized, and they are still viewed as controversial by many. From the public health point of view there has always been, and still is more concern over iron deficiency than over iron excess.

In developed countries, however, few if any people die of iron deficiency, whereas huge numbers may die prematurely and unnecessarily of cancer and vascular disease as a result of iron excess. Our definitions must be changed. Everyone agrees on the definition of *iron deficiency*. In this state, the body does not have adequate iron to satisfy its functional needs, and anemia is the usual symptom. *Iron adequacy* is the desirable state in which all functional needs are met, but excess iron is not present and ferritin stores are very low. *Iron excess* describes the state in which ferritin stores are present, and may range from low to very high, as in the case of hemochromatosis patients. Low iron stores should not be viewed as a liability. Ferritin serves no known function other than to supply iron for accelerated erythropoiesis following a substantial loss of blood. In developed nations, this need has been obviated by our ability to replace lost blood by transfusion. But while the benefit of ferritin has been lost, the risks of ferritin remain, as discussed above.

There is currently no method in common practice that accurately and reliably reflects body iron stores. All of the widely used iron-related methodologies measure parameters that are at best inconsistently correlated with iron stores, and at worst may give very misleading information.

Serum iron concentration reflects almost entirely iron that is bound to transferrin. This measurement alone is not very informative because it fluctuates with time and decreases during inflammation (infection, surgery, malignancy or chronic disease). The fact that the body downregulates its delivery of iron during disease processes is something that we should ponder carefully before deciding to restore this parameter to a "normal" level. Transferrin iron turnover is rapid and represents only a small fraction of total body iron; thus, a low serum iron does not necessarily indicate low stores.

Total iron-binding capacity is a way of measuring transferrin concentration in the serum. It reflects the sum of filled plus empty iron-binding sites. Transferrin concentration rises in iron deficiency (Skikne et al., 1990) but falls during any inflammatory condition. Results therefore have to be carefully interpreted. Also, transferrin is a transport protein and this method measures iron in transit rather than in stores.

Transferrin saturation reflects the percentage of available iron-binding sites that are actually occupied, and may also give some estimation of body iron status. A low transferrin saturation usually indicates iron deficiency but in the case of inflammation or infection a decrease in saturation does not mean that the stores are depleted. In hemochromatosis, a genetic condition of serious iron overload, saturation may reach 100%.

Serum ferritin is one of the better methods to determine iron status in a healthy individual. Using the approximate association that 1 μg/L of serum ferritin is equivalent to 8 mg of stored iron (Walters et al., 1973), iron stores in a person may be estimated. Again, there is strong evidence that serum ferritin, presumably *apoferritin*, in induced by cytokines involved in the inflammatory process (Lee and

Means, 1995; Rogers et al., 1994). Because the assays are immunologically based and measure ferritin protein, not ferritin iron, any measurements conducted during an acute phase response will lead to an overestimation of iron stores. Serum ferritin is, in effect, an acute phase protein that is increased in infection or inflammation, unlike transferrin which decreases. The increase in ferritin concentration is positively correlated with a worsening of the clinical status in severely ill patients (Bobbio-Pallavicini et al., 1989).

Serum ferritin-iron may be a more specific reflection of body iron stores than any of the methods described above for the determination of iron overload. Immunoreactive serum ferritin may be confounded by the rise in serum apoferritin during infection or inflammation, but serum ferritin-iron levels do not change during these conditions. Serum ferritin-iron is determined by immunoprecipitating all serum ferritin, then determining the iron content of this precipitate. While not in widespread use, this method may be the most reliable measurement of body iron stores for clinical use.

Transferrin receptors are located on cell membranes to permit the uptake of the transferrin–iron complex. In iron deficiency, there is an upregulation of transferrin receptors. *Serum transferrin receptor* is presumed to be a soluble product derived from the partial proteolysis of these membrane receptors (Shih et al., 1990). The soluble receptor retains an affinity for transferrin, but only 1% or less of circulating transferrin is bound to a soluble receptor. The concentration of soluble transferrin receptor is thought to be directly proportional to the expression of the membrane receptors. When an individual is in iron excess, serum transferrin receptor concentration is around 6 to 7 mg/L (Skikne et al., 1990). When ferritin is depleted (serum ferritin < 12 μg/L) and functional iron deficiency appears, serum transferrin receptor rises as much as five fold in proportion to the severity of the deficiency. This method appears useful in assessing iron deficiency, but is of no value in assessing iron excess.

IV. IRON AND HEART DISEASE

A state of iron excess is believed by some to be one of the major contributing causes of ischemic heart disease (Sullivan, 1988; McCord, 1991; Salonen et al., 1992b; Van der Kraaij et al., 1988). Sullivan (1989; 1992) has proposed that iron status may be the key to understanding the gender difference with regard to death rates from ischemic heart disease. Serum ferritin levels in men are three or four times those of healthy young women (Jacobs, 1980), presumably reflecting a similar difference in tissue stores of iron.

The Framingham Study revealed that the risk of heart disease in women increased significantly following natural menopause, simple hysterectomy, or hysterectomy with bilateral oophorectomy. This suggests that regular menstrual blood loss (resulting in lower serum and tissue ferritin stores), rather than endogenous estrogen levels, may be the protective factor. There are other studies lending support to this hypothesis (Lauffer, 1991; Lauffer and Sullivan, 1990). A recent study found that serum ferritin is a strong risk factor for acute myocardial infarction (Salonen et al., 1992a) and at levels generally regarded as "normal" (Sullivan, 1989; Sullivan, 1981). Salonen et al. (1992b) have even found that iron deficiency decreases the risk of hypertension and myocardial infarction. Magnusson et al. (1994) were unable to substantiate this finding but nevertheless suggest that transferrin, measured as total iron binding capacity (TIBC), is an independent negative risk factor for myocardial infarction. TIBC increases in iron deficiency (Skikne et al., 1990). Moreover, the serum ferritin results suggest that few subjects in this study were actually iron depleted. Other results do not support the hypothesis that dietary iron in general increases coronary risk in men, but are consistent with an increased risk of myocardial infarction among men with higher intake of heme iron, which is positively associated with iron stores (Ascherio et al., 1994). In addition to the iron status, antioxidant levels appear to be just as important. While iron promotes the initiation of lipid peroxidation, vitamin E terminates lipid peroxidation by reducing the lipid dioxyl radical to a lipid hydroperoxide. Supplementation of vitamin E by at least 100 IU/day was found to decrease the risk of heart disease in both men and women by approximately 40% (Stampfer et al., 1993; Rimm et al., 1993). Population-based studies seeking to correlate iron status with disease should probably therefore take vitamin E status into consideration as a confounding factor. Under normal conditions, the balance between oxidants and antioxidants is a delicate one and it is crucial to maintain this balance. Persons deficient in antioxidants would be further compromised by the contribution of iron to the damaging redox chemistry, resulting in exacerbation of oxidative stress (Figure 1).

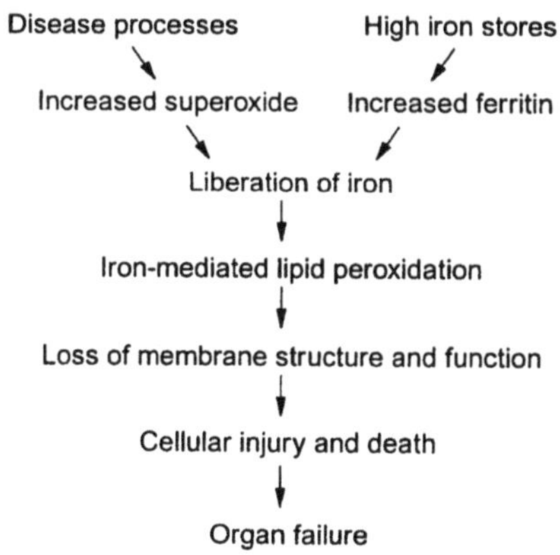

Figure 1. Schematic diagram showing the relationship between superoxide generation resulting from disease processes and the liberation of iron from the storage protein ferritin.

V. ANEMIA OF CHRONIC DISEASE, INFECTION, AND NEOPLASIA

Anemia is often associated with chronic diseases that have include an inflammatory component. This condition has been commonly referred to as "the anemia of chronic disease" but might be more appropriately termed "anemia of chronic inflammation" (Schilling, 1991; Lipschitz, 1990). Serum iron and transferrin (iron binding capacity) levels are low (Cartwright et al., 1946) but normal to increased amounts of iron are found in the bone marrow. Thus, the anemia is not due to iron deficiency, but may rather reflect a problem with the distribution and availability of iron resulting in decreased erythropoiesis. It appears to be more of a hypoferremic response to infection and chronic inflammatory disorders. If it is indeed a deliberate response to inflammation, the evolutionary origins probably derive from the struggle to restrict the availability of iron to invading microbes, even if it means that the host goes through a period of transient anemia. Generally the anemia is described as being mild, with hematocrit levels ranging between 30% to 40% (Cartwright, 1966; Lee, 1983). However, hospitalized patients have been known to have hematocrit values below 25% (Cash and Sears, 1989).

The major pathophysiology of this anemia appears to be decreased production of red blood cells (Sears, 1992) and an impaired ability of the reticuloendothelial cell to recycle the iron derived from previously phagocytized senescent red cells (Lipschitz, 1990). But the pathogenesis of this abnormality remains unclear. The essential difference between iron deficiency anemia and the anemia of chronic disease is the absence of iron stores in the former and normal or increased iron stores in the latter.

The abnormalities in erythropoiesis may be caused by the secretion of certain endogenous pyrogens such as interleukin 1 and tumor necrosis factor by the macrophages. These factors, in addition to being responsible for the fever and inflammation, may also be responsible for the altered ability of the reticuloendothelial system to release iron (Lipschitz, 1990). The suggestion has also been made that the anemia of chronic disease may be mediated by interferon-γ (Means and Krantz, 1992) and that iron restriction is probably a defense against infection and neoplasia (Weinberg, 1992; Weinberg, 1984). Normal human subjects have a mean serum iron concentration of 17.8 μM (range 12–27 μM) and patients with symptomatic diseases have a mean of 5.3 μM (range 2–12 μM) (Cartwright and Lee, 1971). Transferrin iron saturation similarly was lowered from a normal mean of 30% (range 25-50%) to a mean of 15% (range 10–25%). This shift in iron metabolism returns to normal as there is an improvement in the clinical symptoms of the disease. This metabolic shift may be beneficial rather than detrimental. The common belief for several decades that this was indeed a consequence of the disease process led to the practice of treating patients with supplemental doses of iron. Physicians failed to recognize that there were normal to increased levels of iron in the bone marrow and the problem therefore is not one of iron deficiency but rather of a decreased ability to synthesize hemoglobin. Cartwright et al. (1946) noted a rapid disappearance of iron from the blood of patients with chronic infections following the intravenous administration of 1 g ferrous sulfate. There was an increase of 83% in plasma iron in normals within 2–4 h but only a 14% increase in the infected patients. The realization that exogenous iron supplementation may be dangerous by enhancing microbial or neoplastic cell proliferation, or even by amplifying the inflammatory damage accrued by the host, came much later.

Lactoferrin shows broad-spectrum bacteriostatic action that is suppressed by iron but not other transition metals (Oram and Reiter, 1968). Lactoferrin is therefore thought to be a scavenger of iron in septic sites because of its ability to bind this metal at low pH as produced by lactic acid released by microbial cells and/or from the metabolically stimulated leukocytes (Weinberg, 1984; Weinberg, 1986). Lactoferrin is a stronger chelator of iron than transferrin at low pH (Aisen and Leibman, 1972). Lactoferrin may give breast fed infants increased resistance to gastrointestinal infection (Hollan and Johansen, 1993).

Transferrin, like lactoferrin, also exhibits a remarkable broad-spectrum antimicrobial function by its ability to withhold iron from most microbes. As mentioned earlier, however, certain clever microorganisms appear to possess proteolytic enzymes capable of liberating transferrin iron, rendering it up for use by the microbe (Miller and Britigan, 1995).

Malnourished children with kwashiorkor may have extremely low levels of plasma transferrin (3.5 μM, or about 10% of normal), but it is 100% saturated with the metal, even with the low dietary intake (McFarlane et al., 1970). These patients are at greater risk of bacterial and mycotic infections (Weinberg, 1984). Nigerian children with protein-energy malnutrition had elevated serum ferritin levels that ranged between 11–7000 μg/l when compared to normals with a range of 5–107 μg/l (Wickramasinghe et al., 1985). Golden and Ramdath (1987) cite studies showing high iron levels in the liver and bone marrow of children dying from kwashiorkor in Lebanon, India, and South Africa. They have data showing increased levels of plasma ferritin in severely malnourished Jamaican children, particularly in fatal cases. Clearly, iron supplementation in such cases may be detrimental to the patient and may exacerbate clinical symptoms of oxidative stress.

Adult male Zulus have a high incidence of severe hepatic amebiasis associated with excess iron intake resulting from the ingestion of home-brewed beer prepared in iron pots (Diamond et al., 1978). This condition is less common among Zulu women. They are probably exposed similarly to the amoebae but consume lower quantities of beer and their iron status is also lowered by menstruation. In contrast, there was an unusual absence of amebiasis in both males and females in the milk-drinking Masai tribe (infection rate <9%). When their diet was supplemented with iron there was an 83% rise in the rate of infection with an attack of malaria in 17% of the test group while the nonsupplemented control group remained relatively infection free (Murray et al., 1980). Therefore, therapy for patients with the commonly termed anemia of chronic disease must not include oral iron supplementation. Iron is of no apparent benefit in correcting this anemia and might seriously exacerbate any infection. Rather, the underlying infection or inflammation that precipitated this anemia should be managed.

Patients showing symptoms of AIDS were found to have significantly decreased serum iron while showing increased levels of ferritin (presumably apoferritin) (Blumberg et al., 1984). There appears to be an iron-withholding mechanism in AIDS, which of course is characterized by multiple and recurrent infections. In view of the potential undesirable effects of excess iron, supplementation therapy should be considered with great caution.

Critically ill patients often show decreased hemoglobin, serum iron, serum transferrin and transferrin saturation and an increase in serum ferritin levels. In addition, the increase in ferritin concentration is positively correlated with a worsening of the clinical status (Bobbio-Pallavicini et al., 1989).

Chronic inflammation resulting from both infectious and noninfectious sources contribute to carcinogenesis (Ames et al., 1993). An association between body iron stores and the risk of cancer has also been implicated by several studies. Iron status and cancer risk in the first National Health and Nutrition Examination Survey shows increased risk with a decrease in the total iron-binding capacity (TIBC) and a significantly higher transferrin saturation (Stevens et al., 1988). Even though transferrin saturation may not be a good indicator for body iron stores, an increase in TIBC is a good indication of deficiency. That is, iron deficiency may be protective against cancer. These results are consistent with the observation that high body iron stores increase the risk of cancer in men (Selby and Friedman, 1988; Stevens et al., 1986; Stevens et al., 1994). There is also growing evidence from animal studies showing an increase in cancer cell growth with excess iron (Bergeron et al., 1985) and a reduction in the induction of cancer with iron restriction (Thompson et al., 1991; Hann et al., 1988).

Lung cancer rates and mortality have risen to epidemic proportions (Vial, 1986; Davidson et al., 1993; Samet, 1993) and smoking is the single greatest risk factor in this disease. Cigarette smoke and tar can generate a variety of oxidants that can burden not only the lung but the entire system of the smoker (Pryor et al., 1993). The oxygen-derived radicals generated by smoking (Pryor et al., 1993) have been shown to cause DNA single strand breaks (Fielding et al., 1989; Leanderson and Tagesson, 1992) that can be prevented by antioxidants (Leanderson and Tagesson, 1992). Smoking-induced leukocyte adhesion to endothelium can be prevented by superoxide dismutase suggesting a key role of superoxide in this event (Lehr et al., 1993). Alveolar macrophages from smokers contain higher levels of intracellular iron than alveolar macrophages from nonsmokers (Olakanmi et al., 1993; Thompson et al., 1991). Cigarette smoke releases iron from ferritin (Moreno et al., 1992) accompanied by damage to the protein which can be prevented in part by metal chelators and hydroxyl radical scavengers. This mobilization of iron will exacerbate any damage caused by activated phagocytes.

VI. THE ROLE OF IRON IN PARKINSON'S DISEASE

Parkinson's disease is a degenerative disease of the nervous system characterized by involuntary tremulous motion and decreased muscular power. The senses and intellect are not affected. Onset is in middle or late life, with gradual progression and prolonged course. Familial incidence is only 1 or 2%. Iron is thought to contribute to the pathogenesis of Parkinson's disease (Riederer et al., 1989; Youdim et al., 1991; Benshachar et al., 1991; Youdim et al., 1993; Sengstock et al., 1993; Dexter et al., 1991). The association of free-radicals and in particular free iron with Parkinson's disease has been controversial. Increased levels of iron were found in the substantia nigra pars compacta when compared to those of normal control patients (Youdim et al., 1993), but the mechanism is unknown. The nigra may be particularly vulnerable to iron due to its high rate of dopamine oxidation via monoamine oxidase to form hydrogen peroxide (Cohen, 1983; Fahn and Cohen, 1992):

$$\text{Dopamine} + H_2O + O_2 \rightarrow \text{3,4-Dihydroxyphenylacetaldehyde} + NH_3 + H_2O_2$$

Hence, turnover of dopamine is stoichiometrically linked to the generation of hydrogen peroxide which could be associated with the oxyradicals in patients with Parkinson's disease (Adams and Odunze, 1991). Oxidative stress has been associated with nigral cell death (Jenner et al., 1992) and increased levels of lipid peroxidation have been found in substantia nigra in patients with Parkinson's disease (Dexter et al., 1989).

Prior to 1983 there was no animal model of Parkinson's disease, and no mechanistic hypothesis as to its etiology. A bizarre occurrence in the California drug culture provided both when several young heroin abusers developed the clinical symptoms of severe Parkinson's disease (Langston et al., 1983). All had injected a synthetic product, 1-methyl-4-phenyl-1,2,3,6-tetrahydropyridine (MPTP), a by-product of the synthesis of 1-methyl-4-propionoxypyridine, a "designer drug" with heroin-like effects. The compound was found to be oxidized by monoamine oxidases to 1-methyl-4-phenylpyridinium (MPP^+), a potent inhibitor of the NADH dehydrogenase of mitochondrial complex I (Singer and Ramsay, 1990). Hasegawa et al. (1990) found that the MPP^+-poisoned complex I became a site for superoxide production. MPTP can also catalytically generate superoxide radical in an iron-dependent reaction (Poirier and Barbeau, 1985) and can short-circuit mitochondrial electron flow in a manner similar to paraquat (Lambert and Bondy, 1989), thereby producing increased amounts of superoxide and lipid peroxidation. Strong evidence of free radical involvement in the neurotoxic effects of MPTP came from the demonstration that transgenic mice overproducing SOD1 are dramatically resistant to MPTP toxicity (Przedborski et al., 1992). Animal studies have been conducted with exogenous toxins such as 6-hydroxy dopamine (6-OHDA) and MPTP. 6-OHDA selectively destroys nigrostriatal dopamine pathway in the substantia nigra(Hokfelt and Ungerstedt, 1973). 6-OHDA reacts with molecular oxygen to produce superoxide anion radical, hydrogen peroxide and hydroxyl radicals (Heikkila and Cohen, 1973; Cohen and Heikkila, 1974). Hydroxyl radical scavengers have been shown to protect mice in vivo from 6-OHDA toxicity (Cohen et al., 1976). MPTP-treated African green monkeys, which showed obvious contralateral hemiparkinsonism, had significantly elevated iron (Fe^{+2} and Fe^{+3}) in the substantia nigra compacta (Temlett et al., 1994). Three times more iron was found not only in the degenerating dopamine cells themselves but also in the surrounding matrix and glial cells. Similar results were reported by Mochizuki et al. (1994) who have also shown a good correlation of iron accumulation with loss of dopaminergic neurons. These results may reflect the accumulation of iron that has been locally released from ferritin by superoxide.

Ferritin levels are reported to be higher than normal in the substantia nigra in Parkinson's disease (Riederer et al., 1989), as is nonheme iron (Dexter et al., 1989). There have also been reports suggesting that iron in the substantia nigra is bound to melanin (Benshachar et al., 1991) and it is this iron-melanin interaction that could be crucial to the pathogenesis of this disease. Youdim and coworkers (Youdim et al., 1993) have suggested that melanin enhances hydroxyl radical formation by reducing bound Fe^{3+} to Fe^{2+}, which can then reduce hydrogen peroxide to form hydroxyl radical. They also analyzed the substantia nigra iron levels and the caudate nucleus dopamine levels and have shown a direct relationship between the increase of iron and loss of dopamine with the severity of Parkinson's disease. They have speculated that iron may have induced dopaminergic degeneration, or that its accumulation is due to the loss of dopamine neurons or its translocation from other sites due to cellular damage. Population based study of Parkinson's disease shows that it is a late-onset disorder affecting men more frequently than women (Mayeux et al., 1992), consistent with possibility that iron accumulation is a predisposing factor.

As there has been some evidence that the neurodegeneration in the substantia nigra pars compacta of the Parkinson's disease brain may be a consequence of increased iron, Sengstock et al. (1993) have tested this hypothesis by infusing iron (1.25–6.3 nmol) into the rat substantia nigra. The results of their experiments indicate that iron can cause degenerative changes similar to those seen in the disease and further support the hypothesis that iron may indeed contribute to the pathogenesis of Parkinson's disease.

VII. IRON AND AMYOTROPHIC LATERAL SCLEROSIS

Amyotrophic lateral sclerosis (ALS) is a fatally progressive degeneration of the motor neuron system, including precentral motor cortex, corticospinal tracts, the efferent neurons of the anterior horns and/or the bulb (Tandan and Bradley, 1985a; Tandan and Bradley, 1985b; Yoshida et al., 1986). ALS occurs almost twice as frequently in males as in females (Annegers et al., 1991; Kahana and Zilber, 1984; Durrleman and Alperovitch, 1989; Kurtzke, 1991). It occurs most commonly after the age of 50 with a trend towards earlier onset in men than women (Brown et al., 1991). The onset of ALS is unpredictable and the clinical course is very rapid. Death usually occurs within 2–5 years (Gallagher and Sanders, 1987). About 10% of cases are inherited as an autosomal dominant trait, with clinical symptoms appearing after the sixth decade (Mulder et al., 1986; Horton et al., 1976). In most cases, sporadic and autosomal dominant familial ALS (FALS) are clinically similar (Mulder et al., 1986; Swerts and Van Den Bergh, 1976; Husquinet and Franck, 1980).

In some FALS pedigrees the disease is linked to a genetic defect on chromosome 21 (Siddique et al., 1991; Siddique et al., 1989; Takahashi et al., 1994). Recently, Rosen et al. (1993) reported a tight genetic linkage between FALS and the gene that encodes cytosolic Cu,Zn-superoxide dismutase (SOD1). Superoxide dismutase, by preventing the liberation and reduction of iron by superoxide, indirectly controls the rate of iron-dependent lipid peroxidation, a major component of oxidative stress and membrane damage. Many different SOD1 missense mutations have been identified (Rosen et al., 1993). Most if not all of the carriers of these SOD1 mutants have about half the normal erythrocyte SOD activity (Robberecht et al., 1994; Bowling et al., 1995). Many of the mutant proteins are active when expressed in recombinant form, although invariably they are less stable (Borchelt et al., 1994). The decrease in erythrocyte SOD activity is not evident in patients with sporadic ALS, nor in the majority of FALS patients (not possessing mutations in SOD1), suggesting that, if this disease is one of oxidative stress, there may be multiple ways of obtaining the metabolic imbalance.

A transgenic mouse model for ALS has been developed that overexpresses a human ALS mutant SOD1 by five- to eight-fold (Gurney et al., 1994). The mice appear normal for several months, then rapidly develop paralysis in one or more limbs as the result of motor neuron loss from the spinal cord. Because these mice do not possess less than normal SOD activity like ALS patients, but rather possess about four-fold more SOD activity, it has been assumed that the development of ALS-like symptoms is due to some toxic property of the overexpressed mutant SOD protein—some new but unknown "gain of function" (Wong et al., 1995). We believe there is a simpler, more likely explanation. We have shown 'bell-shaped' dose response curves for SOD, reflecting the paradoxical abilities of superoxide radical to both initiate

$$O_2^{\bullet -} + Fe^{+3} \rightarrow O_2 + Fe^{+2}$$
$$Fe^{2+} + LOOH \rightarrow Fe^{3+} + LO\cdot + OH^-$$
$$LO\cdot + LH \rightarrow LOH + L+, \text{etc.}$$

and terminate lipid peroxidation (Nelson et al., 1994):

$$O_2^{\bullet -} + LOO\cdot + H^+ \rightarrow O_2 + LOOH$$

Under any given conditions, the rate of lipid peroxidation occurring reflects the rate of initiation of free radical chains (which is *directly* proportional to superoxide concentration) and the rate of termination of chain reactions (which is *inversely* proportional to superoxide concentration). The practical and perhaps surprising result of this paradox is illustrated in Figure 2. *For any given level of oxidative stress, a unique concentration of SOD exists that minimizes lipid peroxidation.* Contrary to the unlikely hypothesis that all the SOD1 mutant enzymes share a dominant (but unknown) gain of function (Gurney et al., 1994; Bowling et ai., 1995), we believe that ALS is a disease of oxidative imbalance that can be similarly mimicked by either underexpression or overexpression of SOD. Consistent with our view are the observations that decreased SOD causes neuronal death by apoptosis (Rothstein et al., 1994) and the accumulation of lipofuscin granules seen in the motor neurons of patients with ALS (Schiffer et al., 1994). Lipofuscin formation (Sohal et al., 1989; Marzabadi et al., 1990) and apoptosis (Buttke and Sandstrom, 1994) are both clearly related to production of reactive oxygen species.

We have shown that Epstein-Barr virus (EBV) transformed lymphoblasts derived from FALS patients with SOD1 mutations have 24–46% apoptotic cells in culture, whereas those derived from normals have only about 17% apoptosis. The percentage of apoptotic cells in all cultures was decreased to 10–12% in a dose dependent manner by treating with a low molecular weight SOD mimic (Faulkner et al., 1994), illustrating that in these cells oxidative stress is responsible for their increased apoptosis. Conversely, paraquat, the redox-cycling herbicide that increases mitochondrial superoxide production, causes EBV-transformed lymphoblasts to double the percentage of apoptotic cells. This phenomenon is prevented by the ad-

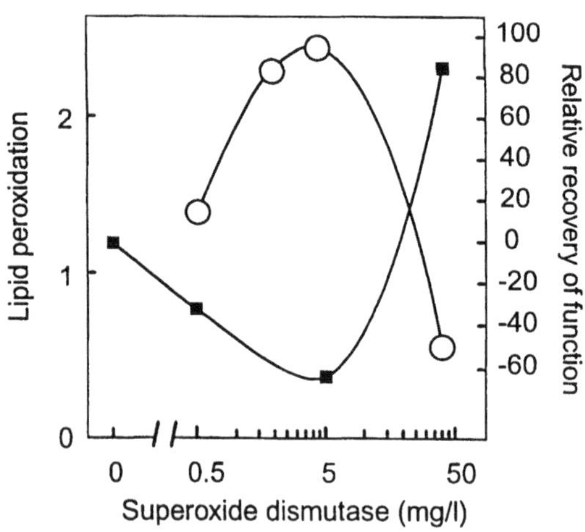

Figure 2. Relationships of net lipid peroxidation and functional recovery of isolated rabbit hearts to administered dosages of superoxide dismutase. Recovery of developed pressure correlates inversely with net lipid peroxidation measured as thiobarbituric acid-reactive substances. Maximal protection is seen at a dosage of 5 mg/L in the coronary perfusate. Data are replotted from Nelson et al. (1994).

dition of mercaptopropionyl glycine, a thiol reducing reagent, to the growth medium. Therefore apoptosis can be induced or prevented by the addition of exogenous oxidants or reductants in cell culture. This opens the door to therapeutic interventions aimed at relieving the oxidative stress associated with FALS SOD1 mutations, and perhaps with all ALS patients.

The loss of 50% of SOD1 activity is not sufficient in itself to cause oxidative death in young healthy cells, as ALS patients show no symptoms for several decades. Oxidative stress, however, is a multifactorial process and certain of its components increase steadily with age. The point at which the disease manifests itself may reflect many genetic, nutritional or environmental factors, all of which contribute to overall oxidative status of the individual. We have proposed that iron accumulation may be one such predisposing or amplifying factor, finally tipping the balance toward lipofuscin accumulation and apoptotic death. Indeed, we find that iron-supplementation of culture medium causes increased apoptosis in EBV-transformed lymphoblasts derived from either normals or FALS patients (unpubished data).

Under normal conditions, the balance between oxidants and antioxidants is a delicate one, and the genetic regulation of SOD appears to be crucial in maintaining this balance. Individuals with the FALS missense mutants, producing SOD with either decreased activity or decreased stability, would be additionally compromised by the contribution of iron to the damaging redox chemistry (See Figure 1.) The reason ALS most commonly occurs after the age of 50 with a trend towards earlier onset in men than women may be because men continue to absorb and store iron throughout their lives, while women are in relative iron-balance until menopause, after which time they too begin to accumulate and store unneeded iron (Cook et al., 1976). There is a striking correlation between iron stores and the gender difference seen in the mortality of ALS (Figure 3). Therefore iron status may be a key to under-

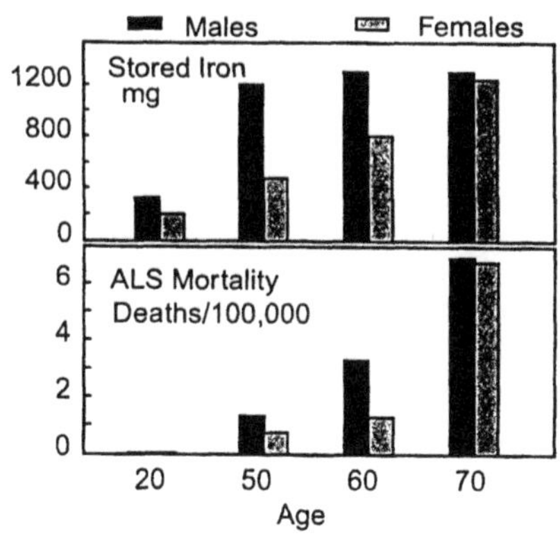

Figure 3. Upper panel shows the gender difference in accumulation of stored iron with age. The data are replotted from Cook et al. (1976). The lower panel shows the gender difference in death rate for amyotrophic lateral sclerosis patients. These data are replotted from Annegers et al. (1991).

standing both the gender difference and the unusual late onset of ALS. This view of ALS as an upset in oxidant/antioxidant balance, which may be the result of a change in any one of a number of components, may also be a useful model for the investigation and reconciliation of non-familial ALS.

VIII. CONCLUSION

In controlled moderation, both free radical formation and the ingestion of iron are biological necessities. In excess, both may contribute to serious damage. Iron depletion due to regular phlebotomy may indeed protect the male population from the various diseases associated with this trace element. A higher survival rate was found for blood donors than for non-blood donors especially between the ages of 50 and 70 (Casale et al., 1983). Several decades ago the incidence of primary liver cancer in Swedish women had increased about three- and-a-half-fold. This increase was associated with the iron fortification of flour which had become a policy in the early 60's, in Sweden (de Sousa and Potaznik, 1984). This type of malignancy is more likely to result from increased dietary intake and/or increased absorption from the gut. We should carefully consider whether an entire population should be supplemented to treat a few who would actually benefit from such fortification. It was once believed that if iron was good for people then more of it was even better. It is time to take a step back and reevaluate iron fortification policies, especially in light of the mounting evidence against excess iron and the role it plays in so many diseases.

REFERENCES

Adams, J.D., Jr., & Odunze, I.N. (1991). Oxygen free radicals and Parkinson's disease. Free Radical Biol. Med. 10, 161-169.

Aisen, P., & Leibman, A. (1972). Lactoferrin and transferrin: A comparative study. Biochim. Biophys. Acta. 257, 314-323.

Ames, B.N., Shigenaga, M.K., & Hagen, T.M. (1993). Oxidants, antioxidants, and the degenerative diseases of aging. Proc. Natl. Acad. Sci. USA 90, 7915-7922.

Annegers, J.F., Appel, S.H., Perkens, P., & Lee, J. (1991). Amyotrophic lateral sclerosis mortality rates in Harris County, Texas. Adv. Neurol., 56, 239-243.

Appelbaum, Y.J., Kuvin, J., Borman, J.B., Uretzky, G., & Chevion, M. (1990). The protective role of neocuproine against cardiac damage in isolated perfused rat hearts. Free Radical Biol. Med., 8, 133-143.

Arroyo, C.M., Kramer, J.H., Dickens, B.F., & Weglicki, W.B. (1987). Identification of free radicals in myocardial ischemia/reperfusion by spin trapping with nitrone DMPO. FEBS Lett. 221, 101-104.

Ascherio, A., Willett, W.C., Rimm, E.B., Giovannucci, E.L., & Stampfer, M.J. (1994). Dietary iron intake and risk of coronary disease among men. Circulation 89, 969-974.

Aust, S.D., Morehouse, L.A., & Thomas, C.E. (1985). Role of metals in oxygen radical reactions. J. Free Radicals Biol. Med. 1, 3-25.

Babbs, C.F. (1985). Role of iron ions in the genesis of reperfusion injury following successful cardiopulmonary resuscitation: preliminary data and a biochemical hypothesis. Ann. Emerg. Med. 14, 777-783.

Badylak, S.F., Simmons, A., Turek, J., & Babbs, C.F. (1987). Protection from reperfusion injury in the isolated rat heart by postischaemic deferoxamine and oxypurinol administration. Cardiovasc. Res. 21, 500-506.

Balla, G., Jacob, H.S., Balla, J., Rosenberg, M., Nath, K., Apple, F., Eaton, J.W., & Vercellotti, G.M. (1992). Ferritin—A cytoprotective antioxidant strategem of endothelium. J. Biol. Chem., 267, 18148-18153.

Bannister, J.V., Bannister, W.H., Hill, H.A.O., & Thornally, P.J. (1982a). Enhanced production of hydroxyl radicals by the xanthine-xanthine oxidase reaction in the presence of lactoferrin. Biochim. Biophys. Acta. 715, 116-120.

Bannister, J.V., Bellavite, P., Davoli, A., Thornally, P.J., & Rossi, F. (1982b). The generation of hydroxyl radicals following superoxide production by neutrophil NADPH oxidase. FEBS Lett. 150, 300-302.

Benshachar, D., Riederer, P., & Youdim, M.B.H. (1991). Iron-melanin interaction and lipid peroxidation—Implications for Parkinson's disease. J. Neurochem. 57, 1609-1614.

Bergeron, R.J., Streiff, R.R., & Elliott, G.T. (1985). Influence of iron on *in vivo* proliferation and lethalityof L1210 cells. J. Nutr. 115, 369-374.

Bernard, M., Menasche, P., Pietri, S., Grousset, C., Piwnica, A., & Cozzone, P.J. (1988). Cardioplegic arrest superimposed on evolving myocardial ischemia: Improved recovery after inhibition of hydroxyl radical generation by peroxidase or deferoxamine. A ^{31}P nuclear resonance study. Circulation, 78 (Suppl. 5) III164-III172.

Bernier, M., Hearse, D.J., & Manning, A.S. (1986). Reperfusion-induced arrhythmias and oxygen-derived free radicals. Studies with "anti-free radical" interventions and a free radical-generating system in the isolated perfused rat heart. Circ. Res. 58, 331-340.

Biemond, P., van Eijk, H.G., Swaak, A.J.G., & Koster, J.F. (1984). Iron mobilization from ferritin by superoxide derived from stimulated polymorphonuclear leukocytes. Possible mechanism in inflammation diseases. J. Clin. Invest. 73, 1576-1579.

Biemond, P., Swaak, A.J., Beindorff, C.M., & Koster, J.F. (1986). Superoxide-dependent and -independent mechanisms of iron mobilization from ferritin by xanthine oxidase. Implications for oxygen-free-radical-induced tissue destruction during ischaemia and inflammation. Biochem. J. 239, 169-173.

Biemond, P., Swaak, A.J., van Eijk, H.G., & Koster, J.F. (1988). Superoxide dependent iron release from ferritin in inflammatory diseases. Free Radical Biol. Med. 4, 185-198.

Blumberg, B.S., Hann, H.W., Mildvan, D., Mathur, U., Lustbader, E., & London, W.T. (1984). Iron and iron binding proteins in persistent generalised lymphadenopathy and AIDS [letter]. Lancet, 1, 347.

Bobbio-Pallavicini, F., Verde, G., Spirano, P., Losi, R., Bosatra, M.G., Braschi, A., Iotti, G., Chiaranda, M., & Villa, S. (1989). Body iron status in critically ill patients: significance of serum ferritin. Intensive Care Med. 15, 171-178.

Bolann, B.J., & Ulvik, R.J. (1987). Release of iron from ferritin by xanthine oxidase. Role of the superoxide radical. Biochem. J., 243, 55-59.

Bolann, B.J., & Ulvik, R.J. (1990). On the limited ability of superoxide to release iron from ferritin. Eur. J. Biochem., 193, 899-904.

Bolli, R., Patel, B.S., Zhu, W.-X., O'Neill, P.G., Hartley, C.J., Charlat, M.L., & Roberts, R. (1987). The iron chelator desferrioxamine attenuates postischemic ventricular dysfunction. Am. J. Physiol. 253(HCP 22), H1372-H1380.

Bolli, R., Patel, B.S., Jeroudi, M.O., Lai, E.K., & McCay, P.B. (1988). Demonstration of free radical generation in "stunned" myocardium of intact dogs with the use of the spin trap α-phenyl *N-tert*-butyl nitrone. J. Clin. Invest. 82, 476-485.

Borchelt, D.R., Lee, M.K., Slunt, H.S., Guarnieri, M., Xu, Z.S., Wong, P.C., Brown, R.H., Price, D.L., Sisodia, S.S., & Cleveland, D.W. (1994). Superoxide dismutase 1 with mutations linked to familial amyotrophic lateral sclerosis possesses significant activity. Proc. Natl. Acad. Sci. USA. 91, 8292-8296.

Bowling, A.C., Barkowski, E.E., Mckennayasek, D., Sapp, P., Horvitz, H.R., Beal, M.F., & Brown, R.H. (1995). Superoxide dismutase concentration and activity in familial amyotrophic lateral sclerosis. J. Neurochem. 64, 2366-2369.

Braughler, J.M., Duncan, L.A., & Chase, R.L. (1986). The involvement of iron in lipid peroxidation. Importance of ferrous ratios in initiation. J. Biol. Chem. 261, 10282-10289.

Brieland, J.K., & Fantone, J.C. (1991). Ferrous iron release from transferrin by human neutrophil-derived superoxide anion: Effect of pH and iron saturation. Arch. Biochem. Biophys. 284, 78-83.

Britigan, B.E., Hassett, D.J., Rosen, G.M., Hamill, D.R., & Cohen, M.S. (1989). Neutrophil degranulation inhibits potential hydroxyl-radical formation. Relative impact of myeloperoxidase and lactoferrin release on hydroxyl-radical production by iron-supplemented neutrophils assessed by spin-trapping techniques. Biochem. J. 264, 447-455.

Brown, R.H.,Jr., Horvitz, H.R., Rouleau, G.A., McKenna-Yasek, D., Beard, C., Sapp, P., Haines, J.L., Gusella, J.F., & Figlewicz, D.A. (1991). Gene linkage in familial amyotrophic lateral sclerosis: a progress report. Adv. Neurol. 56, 215-226.

Buttke, T.M., & Sandstrom, P.A. (1994). Oxidative stress as a mediator of apoptosis. Immunol. Today. 15, 7-10.

Cartwright, G.E., Lauritsen, M.A., Jones, P.J., Merrill, I.M., & Wintrobe, M.M. (1946). The anemia of infection. I. hypoferremia, hypercupremia, and alterations in porphyrin metabolism in patients. J. Clin. Invest. 25, 65-80.

Cartwright, G.E. (1966). The anemia of chronic disorders. Semin. Hematol. 3, 351-375.

Cartwright, G.E., & Lee, G.R. (1971). The anaemia of chronic disorders. Br. J. Haematol. 21, 147-152.

Casale, G., Bignamini, M., & de Nicola, P. (1983). Does blood donation prolong life expectancy? Vox. Sang. 45, 398-399.

Cash, J.M., & Sears, D.A. (1989). The anemia of chronic disease: spectrum of associated diseases in a series of unselected hospitalized patients. Am. J. Med. 87, 638-644.

Cohen, G., Heikkila, R.E., Allis, B., Cabbat, F., Dembiec, D., MacNamee, D., Mytilineou, C., & Winston, B. (1976). Destruction of sympathetic nerve terminals by 6-hydroxydopamine: Protection by 1-phenyl-3-(2-thiazolyl)-2-thiourea, diethyldithiocarbamate, methimazole, cysteamine, ethanol and n-butanol. J. Pharmacol. Exp. Ther. 199, 336-352.

Cohen, G. (1983). The pathobiology of Parkinson's disease: Biochemical aspects of dopamine neuron senescence. J. Neural. Transm [Suppl]., 19, 89-103.

Cohen, G., & Heikkila, R.E. (1974). The generation of hydrogen peroxide, superoxide radical, and hydroxyl radical by 6-hydroxydopamine, dialuric acid, and related cytotoxic agents. J. Biol. Chem. 249, 2447-2452.

Cook, J.D., Finch, C.A., & Smith, N.J. (1976). Evaluation of the iron status of a population. Blood, 48, 449-455.

Davidson, B.J., Hsu, T.C., & Schantz, S.P. (1993). The genetics of tobacco-induced malignancy. Arch. Otolaryngol. Head. Neck. Surg. 119, 1198-1205.

de Sousa, M., & Potaznik, D. (1984). Proteins of the metabolism of iron, cells of the immune system and malignancy. In: Vitamins, Nutrition, and Cancer. (Prasad, K. ed.), pp. 231-239. Karger Basel.

Dexter, D.T., Carter, C.J., Wells, F.R., Javoy-Agid, F., Agid, Y., Lees, A., Jenner, P., & Marsden, C.D. (1989). Basal lipid peroxidation in substantia nigra is increased in Parkinson's disease. J. Neurochem. 52, 381-389.

Dexter, D.T., Carayon, A., Javoy-Agid, F., Agid, Y., Wells, F.R., Daniel, S.E., Lees, A.J., Jenner, P., & Marsden, C.D. (1991). Alterations in the levels of iron, ferritin and other trace metals in Parkinson's disease and other neurodegenerative diseases affecting the basal ganglia. Brain 114, 1953-1975.

Diamond, L.S., Harlow, D.R., Phillips, B.R., & Keister, D.B. (1978). Entamoeba histolytica-iron and nutritional immunity. Arch. Invest. Med. (Mex). i suppl. 1, 329-328.

Durrleman, S., & Alperovitch, A. (1989). Increasing trend of ALS in France and elsewhere: are the changes real? Neurology. 39, 768-773.

Easley, R.M., Schreiner, B.F., & Yu, P.N. (1972). Reversible cardiomyopathy associated with hemochromatosis. N. Engl. J. Med. 287, 866-867.

Fahn, S., & Cohen, G. (1992). The oxidant stress hypothesis in Parkinson's disease—Evidence supporting it. Ann. Neurol. 32, 804-812.

Farber, N.E., Vercellotti, G.M., Jacob, H.S., Pieper, G.M., & Grc·s, G.J. (1988). Evidence for a role of iron-catalyzed oxidants in functional and metabolic stunning in the canine heart. Circ. Res. 63, 351-360.

Faulkner, K.M., Stevens, R.D., & Fridovich, I. (1994). Characterization of Mn(III) complexes of linear and cyclic desferrioxamines as mimics of superoxide dismutase activity. Arch. Biochem. Biophys. 310, 341-346.

Ferrali, M., Ciccoli, L., & Comporti, M. (1989). Allyl alcohol-induced hemolysis and its relation to iron release and lipid peroxidation. Biochem. Pharmacol. 38, 1819-1825.

Ferrali, M., Ciccoli, L., Signorini, C., & Comporti, M. (1990). Iron release and erythrocyte damage in allyl alcohol intoxication in mice. Biochem. Pharmacol. 40, 1485-1490.

Ferrali, M., Signorini, C., Ciccoli, L., & Comporti, M. (1992). Iron release and membrane damage in erythrocytes exposed to oxidizing agents, phenylhydrazine, divicine and isouramil. Biochem. J. 285, 295-301.

Fielding, S., Short, C., Davies, K., Wald, N., Bridges, B.A., & Waters, R. (1989). Studies on the ability of smoke from different types of cigarettes to induce DNA single-strand breaks in cultured human cells. Mutat. Res. 214, 147-151.

Fuller, B.J., Lunec, J., Healing, G., Simpkin, S., & Green, C.J. (1987). Reduction of susceptibility to lipid peroxidation by desferrioxamine in rabbit kidneys subjected to 24-hour cold ischemia and reperfusion. Transplantation 43, 604-606.

Gallagher, J.P., & Sanders, M. (1987). Trauma and amyotrophic lateral sclerosis: A report of 78 patients. Acta Neurol. Scand. 75, 145-150.

Garlick, P.B., Davies, M.J., Hearse, D.J., & Slater, T.F. (1987). Direct detection of free radicals in the reperfused rat heart using electron spin resonance spectroscopy. Circ. Res. 61, 757-760.

Golden, M.H.N., & Ramdath, D. (1987). Free radicals in the pathogenesis of kwashiorkor. Proc. Nutr. Soc., 46, 53-68.

Green, C.J., Healing, G., Simpkin, S., Fuller, B.J., & Lunec, J. (1986). Reduced susceptibility to lipid peroxidation in cold ischemic rabbit kidneys after addition of desferrioxamine, mannitol, or uric acid to the flush solution. Cryobiology 23, 358-365.

Grill, H.P., Zweier, J.L., Kuppusamy, P., Weisfeldt, M.L., & Flaherty, J.T. (1992). Direct measurement of myocardial free radical generation in an *in vivo* model—Effects of postischemic reperfusion and treatment with human recombinant superoxide dismutase. J. Am. Coll. Cardiol. 20, 1604-1611.

Gurney, M.E., Pu, H.F., Chiu, A.Y., Dalcanto, M.C., Polchow, C.Y., Alexander, D.D., Caliendo, J., Hentati, A., Kwon, Y.W., Deng, H.X., Chen, W.J., Zhai, P., Sufit, R.L., & Siddique, T. (1994). Motor neuron degeneration in mice that express a human Cu, Zn superoxide dismutase mutation. Science 264, 1772-1775.

Gutteridge, J.M.C. (1986). Iron promoters of the Fenton reaction and lipid peroxidation can be released from haemoglobin by peroxides. FEBS Lett. 201, 291-295.

Halliwell, B., & Gutteridge, J.M. (1985). The importance of free radicals and catalytic metal ions in human diseases. Mol. Aspects. Med., 8, 89-193.

Halliwell, B., & Gutteridge, J.M.C. (1986). Oxygen free radicals and iron in relation to biology and medicine: some problems and concepts. Arch. Biochem. Biophys. 246, 501-514.

Halliwell, B., & Gutteridge, J.M.C. (1988). Iron as a biological pro-oxidant. ISI Atlas Sci. Biochem. 1, 48-52.

Hann, H.L., Stahlhut, M.W., & Blumberg, B.S. (1988). Iron nutrition and tumor growth:decreased tumor growth in iron-deficient mice. Cancer Res., 48, 4168-4170.

Harris, L.R., Cake, M.H., & Macey, D.J. (1994). Iron release from ferritin and its sensitivity to superoxide ions differs among vertebrates. Biochem. J. 301, 385-389.

Hasegawa, E., Takeshige, K., Oishi, T., Murai, Y., & Minakami, S. (1990). 1-Methyl-4-phenylpyridinium (MPP^+) induces NADH-dependent superoxide formation and enhances NADH-dependent lipid peroxidation in bovine heart submitochondrial particles. Biochem. Biophys. Res. Commun. 170, 1049-1055.

Healing, G., Gower, J., Fuller, B., & Green, C. (1990). Intracellular iron redistribution. An important determinant of reperfusion damage to rabbit kidneys. Biochem. Pharmacol. 39, 1239-1245.

Heikkila, R.E., & Cohen, G. (1973). 6-Hydroxydopamine: Evidence for superoxide radical as an oxidative intermediate. Science, 181, 456-457.

Hobbs, A.J., Fukuto, J.M., & Ignarro, L.J. (1994). Formation of free nitric oxide from L-arginine by nitric oxide synthase: Direct enhancement of generation by superoxide dismutase. Proc. Natl. Acad. Sci. USA 91, 10992-10996.

Hokfelt, T., & Ungerstedt, U. (1973). Specificity of 6-hydroxydopamine induced degeneration of central monoamine neurones: an electron and fluorescence microscopic study with special reference to intracerebral injection on the nigro- striatal dopamine system. Brain Res. 60, 269-297.

Hollan, S., & Johansen, K.S. (1993). Adequate iron stores and the 'nil nocere' principle. Hematologia. 25, 69-84.

Horton, W.A., Eldridge, R., & Brody, J.A. (1976). Familial motor neuron disease. Neurology 26, 460-465.

Hoy, T.G., & Jacobs, A. (1981). Ferritin polymers and the formation of haemosiderin. Br. J. Haematol. 49, 593-602.

Husquinet, H., & Franck, G. (1980). Hereditary amyotrophic lateral sclerosis transmitted for five generations. Clin. Genet. 18, 109-115.

Jacobs, A. (1980). The pathology of iron overload. Academic. Press. ,London., & New York., 427-459.

Jenner, P., Dexter, D.T., Sian, J., Schapira, A.H.V., & Marsden, C.D. (1992). Oxidative stress as a cause of nigral cell death in Parkinson's disease and incidental lewy body disease. Ann. Neurol. 32, S82-S87.

Kahana, E., & Zilber, N. (1984). Changes in the incidence of amyotrophic lateral sclerosis in Israel. Arch. Neurol. 41, 157-160.

Kanner, J., Harel, S., & Granit, R. (1991). Nitric oxide as an antioxidant. Arch. Biochem. Biophys., 289, 130-136.

Krause, G.S., Joyce, K.M., Nayini, N.R., Zonia, C.L., Garritano, A.M., Hoehner, T.J., Evans, A.T., Indreri, R.J., Huang, R.R., Aust, S.D., & White, B.C. (1985). Cardiac arrest and resuscitation: brain iron delocalization during reperfusion. Ann. Emerg. Med. 14, 1037-1043.

Kurtzke, J.F. (1991). Risk factors in amyotrophic lateral sclerosis. Adv. Neurol. 56, 245-270.

Lambert, C.E., & Bondy, S.C. (1989). Effects of MPTP, MPP^+ and paraquat on mitochondrial potential and oxidative stress. Life Sci. 44, 1277-1284.

Langston, J.W., Ballard, P., Tetrud, J.W., & Irwin, I. (1983). Chronic Parkinsonism in humans due to a product of meperidine-analog synthesis. Science 219, 979-980.

Lapenna, D., Degioia, S., Mezzetti, A., Ciofani, G., Consoli, A., Marzio, L., & Cuccurullo, F. (1995). Cigarette smoke, ferritin, and lipid peroxidation. Am. J. Respir. Crit. Care Med. 151, 431-435.

Lauffer, R.B. (1991). Iron stores and the international variation in mortality from coronary artery disease. Med. Hypotheses, 35, 96-102.

Lauffer, R.B., & Sullivan, J.L. (1990). Iron depletion and coronary disease. Am. Heart J. 119, 1448-1449.

Leanderson, P., & Tagesson, C. (1992). Cigarette smoke-induced dna damage in cultured human lung cells -role of hydroxyl radicals and endonuclease activation. Chem. Biol. Interact. 81, 197-208.

Lee, G.R. (1983). The anemia of chronic disease. Semin. Hematol., 20, 61-80.
Lee, M.H., & Means, R.T. (1995). Extremely elevated serum ferritin levels in a university hospital: Associated diseases and clinical significance. Am. J. Med., 98, 566-571.
Lehr, H.A., Kress, E., Menger, M.D., Friedl, H.P., Hubner, C., Arfors, K.E., & Messmer, K. (1993). Cigarette smoke elicits leukocyte adhesion to endothelium in hamsters—inhibition by Cu, Zn-SOD. Free Radical Biol. Med. 14, 573-581.
Lesnefsky, E.J., Repine, J.E., & Horwitz, L.D. (1990). Deferoxamine pretreatment reduces canine infarct size and oxidative injury. J. Pharmacol. Exp. Ther., 253, 1103-1109.
Lipschitz, D.A. (1990). The anemia of chronic disease. J. Am. Ger. Soc., 38, 1258-1264.
Magnusson, M.K., Sigfusson, N., Sigvaldason, H., Johannesson, G.M., Magnusson, S., & Thorgeirsson, G. (1994). Low iron-binding capacity as a risk factor for myocardial infarction. Circulation, 89, 102-108.
Marzabadi, M.R., Sohal, R.S., & Brunk, U.T. (1990). Effect of alpha-tocopherol and some metal chelators on lipofuscin accumulation in cultured neonatal rat cardiac myocytes. Anal. Cell. Pathol., 2, 333-346.
Mayeux, R., Denaro, J., Hemenegildo, N., Marder, K., Tang, M.X., Cote, L.J., & Stern, Y. (1992). A population-based investigation of Parkinson's disease with and without dementia. Relationship to age and gender. Arch. Neurol., 49, 492-497.
McCord, J.M. (1985). Oxygen-derived free radicals in post-ischemic tissue injury. N. Engl. J. Med., 312, 159-163.
McCord, J.M. (1991). Is iron sufficiency a risk factor in ischemic heart disease. Circulation, 83, 1112-1114.
McCord, J.M., & Day, E.D.,Jr. (1978). Superoxide-dependent production of hydroxyl radical catalyzed by iron-EDTA complex. FEBS Lett., 86, 139-142.
McFarlane, H., Reddy, S., Adcock, K.J., Adeshina, H., Cooke, A.R., & Akene, J. (1970). Immunity, transferrin, and survival in kwashiorkor. Br. Med. J. 4, 268-270.
Means, R.T., & Krantz, S.B. (1992). Altered iron metabolism and the anemia of chronic disease. Blood 79, 2797.
Meerson, F.Z., Kagan, V.E., Kozlov, Y.P., Belkina, L.M., & Arkhipenko, Y.V. (1982). The role of lipid peroxidation in pathogenesis of ischemic damage and the antioxidant protection of the heart. Basic Res. Cardiol. 77, 465-485.
Menasche, P., Grousset, C., Mouas, C., & Piwnica, A. (1990). A promising approach for improving the recovery of heart transplants: Prevention of free radical injury through iron chelation by deferoxamine. J. Thorac. Cardiovasc. Surg. 100, 13-21.
Miller, R.A., & Britigan, B.E. (1995). Protease-cleaved iron-transferrin augments oxidant- mediated endothelial cell injury via hydroxyl radical formation. J. Clin. Invest., 95, 2491-2500.
Milman, N., Sengelov, H., & Dombernowsky, P. (1991). Iron status markers in patients with small cell carcinoma of the lung - relation to survival. Br. J. Cancer., 64, 895-898.
Mochizuki, H., Imai, H., Endo, K., Yokomizo, K., Murata, Y., Hattori, N., & Mizuno, Y. (1994). Iron accumulation in the substantia nigra of 1-methyl-4- phenyl-1,2,3,6-tetrahydropyridine (MPTP)-induced hemiparkinsonian monkeys. Neurosci. Lett., 168, 251-253.
Monteiro, H.P., Vile, G.F., & Winterbourn, C.C. (1989). Release of iron from ferritin by semiquinone, anthracycline, bipyridyl, and nitroaromatic radicals. Free Radical Biol. Med. 6, 587-591.
Monteiro, H.P., & Winterbourn, C.C. (1989). 6-Hydroxydopamine releases iron from ferritin and promotes ferritin-dependent lipid peroxidation. Biochem. Pharmacol., 38, 4177-4182.
Moreno, J.J., Foroozesh, M., Church, D.F., & Pryor, W.A. (1992). Release of iron from ferritin by aqueous extracts of cigarette smoke. Chem. Res. Toxicol., 5, 116-123.
Moss, M., Kurzner, S., Razlog, Y., & Lister, G. (1988). Hypoxanthine and lactate concentrations in lambs during hypoxic and stagnant hypoxia. Am. J. Physiol. 255, H53-H59.
Mulder, D.W., Kurland, L.T., Offord, K.P., & Beard, C.M. (1986). Familial adult motor neuron disease: amyotrophic lateral sclerosis. Neurology., 36, 511-517.

Murphy, M.E., & Sies, H. (1991). Reversible conversion of nitroxyl anion to nitric oxide by superoxide dismutase. Proc. Natl. Acad. Sci. USA 88, 10860-10864.

Murray, M.J., Murray, A., & Murray, C.J. (1980). The salutary effect of milk on amoebiasis and its reversal by iron. Br. Med. J. 1351-1352.

Myers, C.L., Weiss, S.J., Kirsh, M.M., & Shlafer, M. (1985). Involvement of hydrogen peroxide and hydroxyl radical in the 'oxygen paradox': reduction of creatine kinase release by catalase, allopurinol or deferoxamine, but not by superoxide dismutase. J. Mol. Cell. Cardiol. 17, 675-684.

Nelson, S.K., Bose, S.K., & McCord, J.M. (1994). The toxicity of high-dose superoxide dismutase suggests that superoxide can both initiate and terminate lipid peroxidation in the reperfused heart. Free Radical Biol. Med. 16, 195-200.

Norskov-Lauritsen, N., Demant, E.J., & Ebbesen, P. (1987). Mobilization of ferritin-iron by the anticancer agent bleomycin. Biochem. Pharmacol. 36, 2685-2686.

O'Connell, M., Halliwell, B., Moorhouse, C.P., Aruoma, O.I., Baum, H., & Peters, T.J. (1986). Formation of hydroxyl radicals in the presence of ferritin and haemosiderin. Is haemosiderin formation a biological protective mechanism? Biochem. J. 234, 727-731.

O'Connell, M.J., Baum, H., & Peters, T.J. (1986). Haemosiderin-like properties of free-radical-modified ferritin. Biochem. J., 240, 297-300.

Olakanmi, O., McGowan, S.E., Hayek, M.B., & Britigan, B.E. (1993). Iron sequestration by macrophages decreases the potential for extracellular hydroxyl radical formation. J. Clin. Invest. 91, 889-899.

Omar, R., Nomikos, I., Piccorelli, G., Savino, J., & Agarwal, N. (1989). Prevention of postischaemic lipid peroxidation and liver cell injury by iron chelation. Gut 30, 510-514.

Oram, J.D., & Reiter, B. (1968). Inhibition of bacteria by lactoferrin and other iron-chelating agents. Biochim. Biophys. Acta, 170, 351-365.

Paller, M.S., & Hedlund, B.E. (1988). Role of iron in postischemic renal injury in the rat. Kidney. Int. 34, 474-480.

Poirier, J., & Barbeau, A. (1985). A catalyst function for MPTP in superoxide formation. Biochem. Biophys. Res. Commun., 131, 1284-1289.

Powers, R.H., Stadnicka, A., Kalbfleish, J.H., & Skibba, J.L. (1992). Involvement of xanthine oxidase in oxidative stress and iron release during hyperthermic rat liver perfusion. Cancer Res. 52, 1699-1703.

Pryor, W.A., Stone, K., Cross, C.E., Machlin, L., & Packer, L. (1993). Oxidants in cigarette smoke—radicals, hydrogen peroxide, peroxynitrate, and peroxynitrite. Tobacco. Smoking., & Nutrition. 686.

Przedborski, S., Kostic, V., Jacksonlewis, V., Naini, A.B., Simonetti, S., Fahn, S., Carlson, E., Epstein, C.J., & Cadet, J.L. (1992). Transgenic mice with increased Cu/Zn-superoxide dismutase activity are resistant to N-methyl-4-phenyl-1,2,3,6-tetrahydropyridine-induced neurotoxicity. J. Neurosci. 12, 1658-1667.

Reddy, B.R., Kloner, R.A., & Przyklenk, K. (1989). Early treatment with deferoxamine limits myocardial ischemic/reperfusion injury. Free Radical Biol. Med. 7, 45-52.

Reif, D.W., Schubert, J., & Aust, S.D. (1988). Iron release from ferritin and lipid peroxidation by radiolytically generated reducing radicals. Arch. Biochem. Biophys. 264, 238-243.

Reif, D.W. (1992). Ferritin as a source of iron for oxidative damage. Free Radical Biol. Med. 12, 417-427.

Reif, D.W., & Simmons, R.D. (1990). Nitric oxide mediates iron release from ferritin. Arch. Biochem. Biophys. 283, 537-541.

Riederer, P., Sofic, E., Rausch, W.-D., Schmidt, B., Reynolds, G.P., Jellinger, K., & Youdim, M.B.H. (1989). Transition metals, ferritin, glutathione, and ascorbic acid in Parkinsonian brains. J. Neurochem. 52, 515-520.

Rimm, E.B., Stampfer, M.J., Ascherio, A., Giovannucci, E., Colditz, G.A., & Willett, W.C. (1993). Vitamin-e consumption and the risk of coronary heart disease in men. N. Engl. J. Med. 328, 1450-1456.

Robberecht, W., Sapp, P., Viaene, M.K., Rosen, D., McKenna-Yasek, D., Haines, J., Horvitz, R., Theys, P., & Brown, R.,Jr. (1994). Cu/Zn superoxide dismutase activity in familial and sporadic amyotrophic lateral sclerosis. J. Neurochem. 62, 384-387.

Rogers, J., Lacroix, L., Durmowitz, G., Kasschau, K., &riotakis, J., & Bridges, K.R. (1994). The role of cytokines in the regulation of ferritin expression. Adv. Exp. Med. Biol. 356, 127-132.

Rosen, D.R., Siddique, T., Patterson, D., Figlewicz, D.A., Sapp, P., Hentati, A., Donaldson, D., Goto, J., O'Regan, J.P., Deng, H.X., & et al, (1993). Mutations in Cu/Zn superoxide dismutase gene are associated with familial amyotrophic lateral sclerosis [published erratum appears in Nature 1993 Jul 22;364(6435):362] [see comments]. Nature, 362, 59-62.

Rothman, R.J., Serroni, A., & Farber, J.L. (1992). Cellular pool of transient ferric iron, chelatable by deferoxamine and distinct from ferritin, that is involved in oxidative cell injury. Mol. Pharmacol. 42, 703-710.

Rothstein, J.D., Bristol, L.A., Hosler, B., Brown, R.H., & Kuncl, R.W. (1994). Chronic inhibition of superoxide dismutase produces apoptotic death of spinal neurons. Proc. Natl. Acad. Sci. USA 91, 4155-4159.

Rubanyi, G.M., & Vanhoutte, P.M. (1986). Oxygen-derived free radicals, endothelium, and responsiveness of vascular smooth muscle. Am. J. Physiol., 250, H815-H821.

Salonen, J.T., Nyyssonen, K., Korpela, H., Tuomilehto, J., Seppanen, R., & Salonen, R. (1992a). High stored iron levels are associated with excess risk of myocardial infarction in eastern Finnish men. Circulation, 86, 803-811.

Salonen, J.T., Salonen, R., Nyyssonen, K., & Korpela, H. (1992b). Iron sufficiency is associated with hypertension and excess risk of myocardial infarction: The Kuopio Ischaemic Heart Disease Risk Factor Study (KIHD). Circulation 85, 864.

Samet, J.M. (1993). The epidemiology of lung cancer. Chest, 103, 20S-29S.

Schiffer, D., Attanasio, A., Chio, A., Migheli, A., & Pezzulo, T. (1994). Ubiquitinated dystrophic neurites suggest corticospinal derangement in patients with amyotrophic lateral sclerosis. Neurosci. Lett. 180, 21-24.

Schilling, R.F. (1991). Anemia of chronic disease: a misnomer. Ann. Int. Med. 115, 572-573.

Sears, D.A. (1992). Anemia of chronic disease. Med. Clin. North Am. 76, 567-579.

Selby, J.V., & Friedman, G.D. (1988). Epidemiologic evidence of an association between body iron stores and risk of cancer. Int. J. Cancer 41, 677-682.

Sengstock, G.J., Olanow, C.W., Menzies, R.A., Dunn, A.J., & Arendash, G.W. (1993). Infusion of iron into the rat substantia-nigra - nigral pathology and dose-dependent loss of striatal dopaminergic markers. J. Neurosci. Res., 35, 67-82.

Shih, Y.J., Baynes, R.D., Hudson, B.G., Flowers, C.H., Skikne, B.S., & Cook, J.D. (1990). Serum transferrin receptor is a truncated form of tissue receptor. J. Biol. Chem., 265, 19077-19081.

Siddique, T., Pericak-Vance, M.A., Brooks, B.R., Roos, R.P., Hung, W.Y., Antel, J.P., Munsat, T.L., Phillips, K., Warner, K., Speer, M., & et al, (1989). Linkage analysis in familial amyotrophic lateral sclerosis. Neurology. 39, 919-925.

Siddique, T., Figlewicz, D.A., Pericak-Vance, M.A., Haines, J.L., Rouleau, G., Jeffers, A.J,, Sapp, P., Hung, W.Y., Bebout, J., McKenna-Yasek, D., Deng, G., Horvitz, H.R., Gusella, J.F., Brown, R.H.,Jr., & Roses, A.D. (1991). Linkage of a gene causing familial amyotrophic lateral sclerosis to chromosome 21 and evidence of genetic-locus heterogeneity. N. Engl. J. Med. 324, 1381-1384.

Singer, T.P., & Ramsay, R.R. (1990). Mechanism of the neurotoxicity of MPTP. An update. FEBS Lett., 274, 1-8.

Skikne, B.S., Flowers, C.H., & Cook, J.D. (1990). Serum transferrin receptor: a quantitative measure of tissue iron deficiency. Blood 75, 1870-1876.

Sohal, R.S., Marzabadi, M.R., Galaris, D., & Brunk, U.T. (1989). Effect of ambient oxygen concentration on lipofuscin accumulation in cultured rat heart myocytes—A novel *in vitro* model of lipofuscinogenesis. Free Radical Biol. Med., 6, 23-30.

Sohal, R.S., & Brunk, U.T. (1990). Lipofuscin as an indicator of oxidative stress and aging. Adv. Exp. Med. Biol., 266, 17-30.

Stampfer, M.J., Hennekens, C.H., Manson, J.E., Colditz, G.A., Rosner, B., & Willett, W.C. (1993). Vitamin-e consumption and the risk of coronary disease in women. N. Engl. J. Med., 328, 1444-1449.

Stevens, R.G., Beasley, R.P., & Blumberg, B.S. (1986). Iron-binding proteins and risk of cancer in Taiwan. J. Nat. Cancer Inst., 76, 605-610.

Stevens, R.G., Jones, D.Y., Micozzi, M.S., & Taylor, P.R. (1988). Body iron stores and the risk of cancer. N. Engl. J. Med., 319, 1047-1052.

Stevens, R.G., Graubard, B.I., Micozzi, M.S., Nerhshi, K., & Blumberg, B.S. (1994). Moderate elevation of body iron level and increased risk of cancer occurrence and death. Int. J. Cancer 56, 364-369.

Sullivan, J.L. (1981). Iron and the sex difference in heart disease risk. Lancet 1, 1293-1294.

Sullivan, J.L. (1988). Iron, plasma antioxidants, and the 'oxygen radical disease of prematurity'. Am. J. Dis. Child. 142, 1341-1344.

Sullivan, J.L. (1989). The iron paradigm of ischemic heart disease. Am. Heart J., 117, 1177-1188.

Sullivan, J.L. (1992). Stored iron and ischemic heart disease—empirical support for a new paradigm. Circulation, 86, 1036-1037.

Swerts, L., & Van Den Bergh, R. (1976). [Familial amytrophic lateral sclerosis. A study of a family suffering from this disease for three generations]. J. Genet. Hum. 24, 247-255.

Takahashi, H., Makifuchi, T., Nakano, R., Sato, S., Inuzuka, T., Sakimura, K., Mishina, M., Honma, Y., Tsuji, S., & Ikuta, F. (1994). Familial amyotrophic lateral sclerosis with a mutation in the Cu/Zn superoxide dismutase gene. Acta Neuropathol., 88, 185-188.

Tandan, R., & Bradley, W.G. (1985a). Amyotrophic lateral sclerosis: Part 2. Etiopathogenesis. Ann. Neurol., 18, 419-431.

Tandan, R., & Bradley, W.G. (1985b). Amyotrophic lateral sclerosis: Part 1. Clinical features, pathology, and ethical issues in management. Ann. Neurol., 18, 271-280.

Tavazzi, B., Lazzarino, G., Dipierro, D., & Giardina, B. (1992). Malondialdehyde production and ascorbate decrease are associated to the reperfusion of the isolated postischemic rat heart. Free Radical Biol. Med. 13, 75-78.

Temlett, J.A., Landsberg, J.P., Watt, F., & Grime, G.W. (1994). Increased iron in the substantia nigra compacta of the MPTP-lesioned hemiparkinsonian African green monkey: evidence from proton microprobe elemental microanalysis. J. Neurochem. 62, 134-146.

Thomas, C.E., Morehouse, L.A., & Aust, S.D. (1985). Ferritin and superoxide-dependent lipid peroxidation. J. Biol. Chem., 260, 3275-3280.

Thomas, C.E., & Aust, S.D. (1986). Reductive release of iron from ferritin by cation free radicals of paraquat and other bipyridyls. J. Biol. Chem., 261, 13064-13070.

Thompson, A.B., Bohling, T., Heires, A., Linder, J., & Rennard, S.I. (1991). Lower respiratory tract iron burden is increased in association with cigarette smoking. J. Lab. Clin. Med., 117, 493-499.

Thompson, H.J., Kennedy, K., Witt, M., & Juzefyk, J. (1991). Effect of dietary iron deficiency or excess on the induction of mammary carcinogenesis by l-methyl-l-nitrosourea. Carcinogenesis, 12, 111-114.

Van der Kraaij, A.M.M., Mostert, L.J., van Eijk, H.G., & Koster, J.F. (1988). Iron-load increases the susceptibility of rat hearts to oxygen reperfusion damage: Protection by the antioxidant (+)-cyanidanol-3 and deferoxamine. Circulation, 78, 442-449.

Vial, W.C. (1986). Cigarette smoking and lung disease. Am. J. Med. Sci. 291, 130-142.

Voogd, A., Sluiter, W., van Eijk, H.G., & Koster, J.F. (1992). Low molecular weight iron and the oxygen paradox in isolated rat hearts. J. Clin. Invest. 90, 2050-2055.

Voogd, A. (1993). Iron and the Oxygen Paradox in Ischemic Hearts (Den Haag, Netherlands: CIP-DATA Koninklijke Bibliotheek).

Walters, G.O., Miller, F.M., & Worwood, M. (1973). Serum ferritin concentration and iron stores in normal subjects. J. Clin. Path. 26, 770-772.

Weinberg, E.D. (1984). Iron withholding: a defense against infection and neoplasia. Am. Physiol. Soc., 64, 65-101.

Weinberg, E.D. (1986). Iron, infection, and neoplasia. Clin. Physiol. Biochem. 4, 50-60.

Weinberg, E.D. (1992). Iron depletion—a defense against intracellular infection and neoplasia. Life Sci., 50, 1289-1297.

White, B.C., Krause, G.S., Aust, S.D., & Eyster, G.E. (1985). Postischemic tissue injury by iron-mediated free radical lipid peroxidation. Ann. Emerg. Med. 14, 804-809.

Wickramasinghe, S.N., Gill, D.S., Broom, G.N., Akinyanju, O.O., & Grange, A. (1985). Limited value of serum ferritin in evaluating iron status in children with protein-energy malnutrition. Scand. J. Haematol. 35, 292-298.

Williams, R.E., Zweier, J.L., & Flaherty, J.T. (1991). Treatment with deferoxamine during ischemia improves functional and metabolic recovery and reduces reperfusion-induced oxygen radical generation in rabbit hearts. Circulation, 83, 1006-1014.

Wong, P.C., Pardo, C.A., Borchelt, D.R., Lee, M.K., Copeland, N.G., Jenkins, N.A., Sisodia, S.S., Cleveland, D.W., & Price, D.L. (1995). An adverse property of a familial ALS-linked SOD1 mutation causes motor neuron disease characterized by vacuolar degeneration of mitochondria. Neuron 14, 1105-1116.

Yamashita, C., Oobo, H., Tsuji, F., Tobe, S., Yamamoto, H., Nakamura, H., Okada, M., & Nakamura, K. (1992). Effect of prostaglandin-I_2 and superoxide dismutase on reperfusion injury of warm ischemic lung. Ann. Thorac. Surg. 54, 921-924.

Yoshida, S., Mulder, D.W., Kurland, L.T., Chu, C.P., & Okazaki, H. (1986). Follow-up study on amyotrophic lateral sclerosis in Rochester, Minn., 1925 through 1984. Neuroepidemiology., 5, 61-70.

Youdim, M.B.H., Ben-Shachar, D., & Riederer, P. (1991). Iron in brain function and dysfunction with emphasis on Parkinson's disease. Eur. Neurol., 31 Suppl. 1, 34-40.

Youdim, M.B.H., Benshachar, D., & Riederer, P. (1993). The possible role of iron in the etiopathology of Parkinson's disease—review. Movement. Disord. 8, 1-12.

INDEX

J A I P R E S S

Advances in Molecular and Cell Biology

Edited by **E. Edward Bittar**, *Department of Physiology, University of Wisconsin, Madison*

Cell biology is a rapidly developing discipline which brings together many of the branches of the biological sciences that were separate in the past. The interrelations between cell structure and function at the molecular and subcellular level are the central connecting theme of this series. "As the twentieth century nears its close," writes Kenneth Miller, "the development of an enormous range of tools and techniques, some physical, some chemical, some biological, has changed the situation forever. Cell biology today crosses the boundary, links the molecule with the organelle, associates the cellular response with the larger organism." Most of the contributors are cell and molecular biologists who are currently engaged in research and have made important contributions to our present understanding of cell function.

Volume 24, Cell Structure and Signaling
1997, 272 pp. $109.50/£69.50
ISBN 0-7623-0288-7

Edited by **Robert H. Getzenberg**, *Departments of Pathology, Surgery, Medicine, and Pharmacology, University of Pittsburgh*

CONTENTS: Preface, *Robert H. Getzenberg.* Extracellular Matrix and Nuclear Matrix Interactions May Regulate Apoptosis and Tissue-Specific Gene Expression: A Concept Whose Time Has Come, *Sophie Lelièvre, Valerie M. Weaver, Carolyn A. Larabell, and Mina J. Bissell.* Role of the Extracellular Matrix and Cytoskeleton in the Regulation of Cyclins, Cyclin-Dependent Kinase Inhibitors, and Anchorage-Dependent Growth, *Richard K. Assoian, Xiaoyun Zhu, Gabriela E. Davey, and Maria Elena Bottazzi.* Aptamer Adaptability: Utilizing Tumor Cell Surface Heterogeneity to Self-Select Appropriate Diagnostic and Therapeutic Agents, *Miguel Garcia Schürmann and Donald S. Coffey.* Cytoskeleton-Mediated Aspects of Signal Transduction, *Raquell M. Holmes, Mary Jo Carabatsos, and David F. Albertini.* The Role of the Cytoskeleton in Adhesion-Mediated Signaling and Gene Expression, *Avri Ben-Ze'ev and Alexander D. Bershadsky.* Subnuclear Trafficking of Steroid Receptors, *Donald B. DeFranco, Jimin Liu, Yuting Tang, and Jun Yang.* The Role of Nuclear Matrix in Tissue-Specific Gene Expression, *Michael J. Horton and Robert H. Getzenberg.* Explaining Aberrations of Cell Structure and Cell Signaling in Cancer Using Complex Adaptive Systems, *Eric D. Schwab and Kenneth J. Pienta.* Index.

Also Available:

Volumes 1-14, 16-21, 24 (1987-1997) $109.50/£69.50 each
Volumes 15, 22-23 (2 part sets) $219.00/£$139.00 each

www.ingramcontent.com/pod-product-compliance
Ingram Content Group UK Ltd.
Pitfield, Milton Keynes, MK11 3LW, UK
UKHW020818290926
15470UKWH00003B/1227